IMAGES
of America

SPRINGFIELD FIREFIGHTING

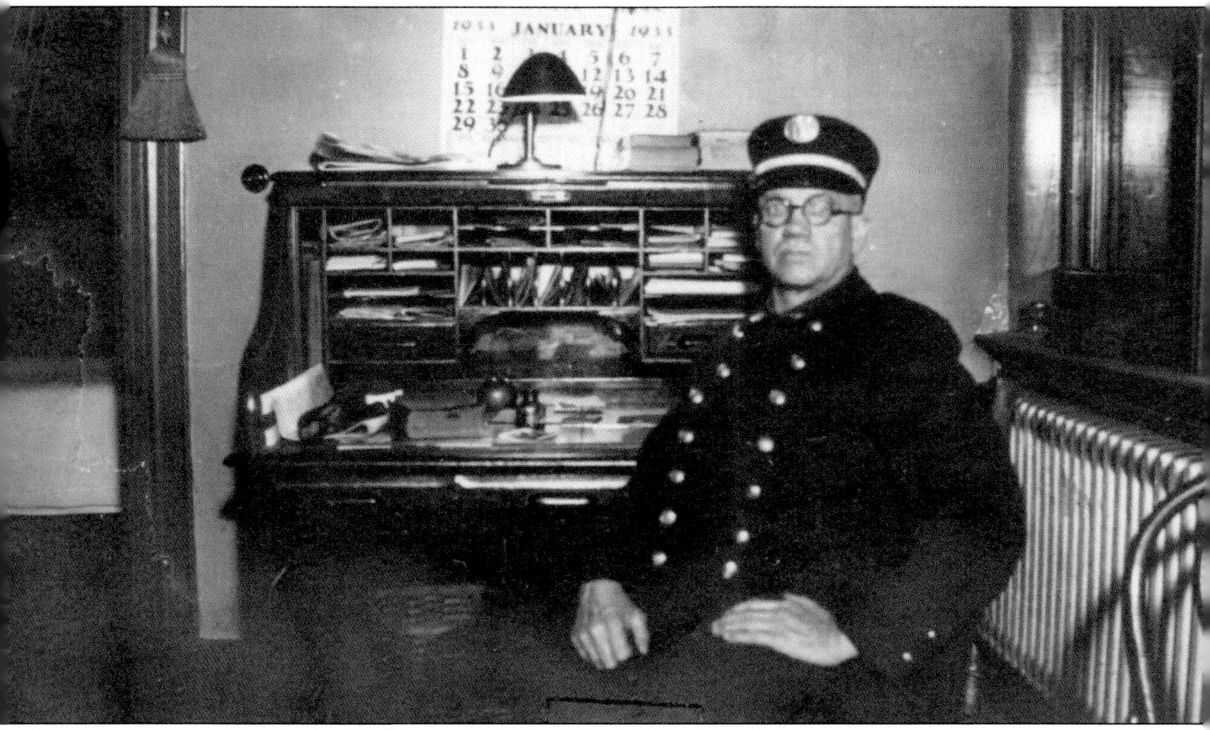

Noah Pooler was born in Vasselboro, Maine, in 1877 and served in the U.S. Army during the Spanish American War. He was later stationed at the Springfield Armory, and when his enlistment was over he stayed in Springfield and joined the Springfield Fire Department in 1907. He was promoted to lieutenant in 1908 and then to captain in 1920. He retired from the department in 1942. He passed away on January 1, 1956, in his home in Springfield. (Courtesy Noah Pooler family archives.)

On the cover: Please see page 101. (Courtesy Noah Pooler family archives.)

IMAGES
of America

SPRINGFIELD FIREFIGHTING

Bert D. Johanson and Nancy B. Johanson assisted by Fred Rodriquez

ARCADIA
PUBLISHING

Copyright © 2007 by Bert D. Johanson and Nancy B. Johanson assisted by Fred Rodriquez
ISBN 978-0-7385-5011-4

Published by Arcadia Publishing
Charleston SC, Chicago IL, Portsmouth NH, San Francisco CA

Printed in the United States of America

Library of Congress Catalog Card Number: 2006940916

For all general information contact Arcadia Publishing at:
Telephone 843-853-2070
Fax 843-853-0044
E-mail sales@arcadiapublishing.com
For customer service and orders:
Toll-Free 1-888-313-2665

Visit us on the Internet at www.arcadiapublishing.com

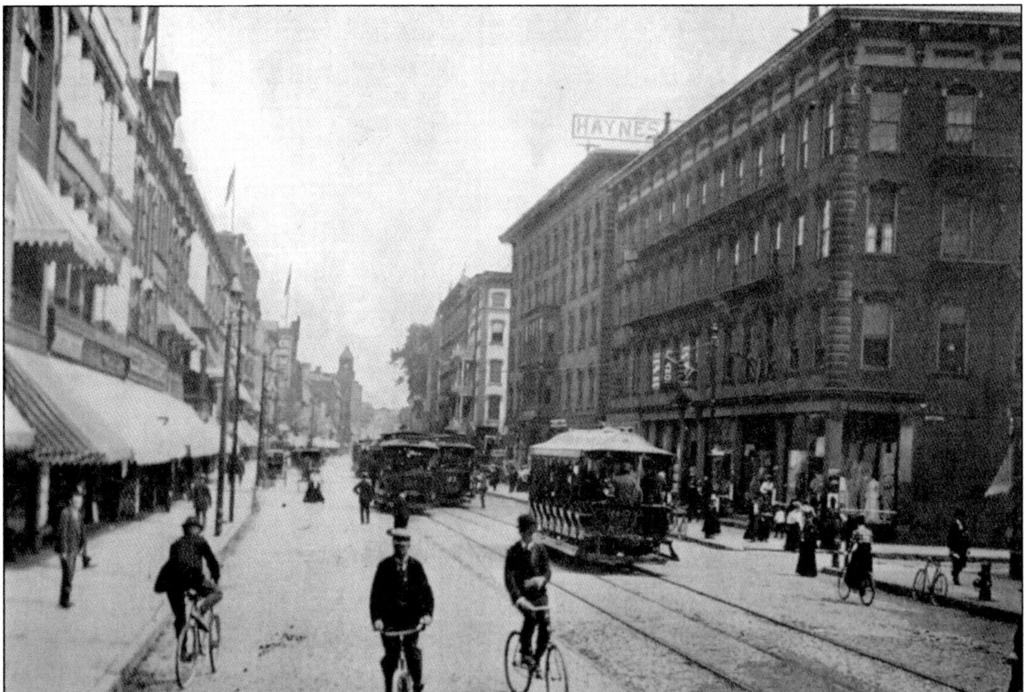

This is a photograph of Main Street in Springfield in 1900, before automobiles took over the streets of the cities in America. During this era, streetcars, wagons, and bicycles were the primary vehicles on the road. In 1900, there were only about 1,500 cars in the state of Massachusetts. (Courtesy Connecticut Electric Railway Association.)

Contents

ACKNOWLEDGMENTS

This book could not have been possible without the help of the Pooler family of Springfield. Noah Pooler became part of the Springfield Fire Department in 1907 and retired as a captain in 1942. He took many of the pictures seen in this book, and thanks to his son George Pooler and grandson William Pooler, these pictures have been preserved for future generations to see and appreciate. History might have slipped from our grasp if it was not for their efforts. Unfortunately, it has been impossible to identify many of the firefighters who appear in this book, but this work is a tribute to their dedication in the performance of the duty they loved.

Special thanks to Fred Rodriguez of the Springfield Fire Department who was an invaluable source, providing information for many of the pictures in this book. Rodriguez has 29 years of service on the department as a firefighter, pump operator, dispatcher, and fire inspector and is probably the most knowledgeable person on the history of the Springfield Fire Department today.

We would also like to thank Geoffery Neilsen of Wethersfield, Connecticut, and Raymond Pond of Agawam for their contributions of photographs and historical information.

We would also like to extend our appreciation to Maggie Humberston, head of library and archives of the Springfield museums, who allowed us to research the history of the Springfield Fire Department using their extensive archives.

Bert and Nancy Johanson are both longtime members of the Connecticut Fire Museum in East Windsor, Connecticut, and have written a number of books on the history of street railways and firefighting in Connecticut. The museum was founded in 1968 by a group of members at the Connecticut Trolley Museum in East Windsor, Connecticut, who realized that they had a common interest in fire engines as well as streetcars. After acquiring a number of older fire apparatus, the group formed the Connecticut Fire Museum. The museum was opened in 1975, and, in conjunction with the Connecticut Trolley Museum, has over 25,000 visitors a year. Both organizations are living history museums where both antique fire apparatus and streetcars, some nearly 100 years old, can be seen in operation. All of the author's proceeds from this book will be donated to the Connecticut Fire Museum.

Connecticut Fire Museum
58 North Road
East Windsor, CT 06088
Telephone: (860) 623-4732
www.ct-trolley/Firemuseum

INTRODUCTION

The history of the Springfield Fire Department dates back to 1792 when the first hand engine, the *Lion*, was purchased from a company in Philadelphia. The *Lion* pumped water by having 20 men push up and down on the brakes, which were handles located on each end of the engine. This engine did not have a suction hose, so it was necessary to have the water poured into the troughs on each side. It required about 100 people, both men and women, to pass buckets from the source of water. The *Lion* was housed in the first fire station in Springfield in a building located at the center of Springfield's settlement on State and Market Streets.

On January 17, 1794, a fire club was organized and the first annual meeting was held in the home of Zenos Parson. There were 50 members in the club, all of whom were prominent citizens. The regulations of this primitive club were few and simple; the only stringent law being one that required each member to keep in his house two buckets for water and two sacks for saving valuables from burning buildings. In this era, buildings that caught fire usually burned down and the firefighters worked to save the surrounding structures. In the days of the *Lion*, the fire wardens, as the fire chiefs were called in this era, directed the men at the scene. They carried long brass-tipped poles, which they used to jab those working on the brakes to make them work faster.

The next engine purchased for use in Springfield was the *Tiger*, a hand pumper like the *Lion*. This engine was soon exchanged for a more modern engine called the *Cataract*. The *Lion* was last used in 1840 but was not discarded until after 1865, when the city began the conversion to steam pumpers.

In 1830, the Massachusetts legislature passed an act establishing fire districts, and three years later, Springfield organized a fire department. Like the members of its predecessor, the fire club, the new volunteer fire department consisted of prominent members of the community.

In 1844, the city purchased a Waterman hand engine that was called *Niagara*, the first engine of its type in the Connecticut Valley. With this new engine, the department was able to be very effective at fighting fires. Also in 1844, Elisha Blake, then 96 years old, resigned as chief engineer. Having entered the service in 1809, he served not only as chief, but was also a bucket man, hose man, and fire warden. In that same year, the Western Railroad Company, which later became the Boston and Albany (now CSX), purchased a hand engine called the *Ocean* to protect its car and locomotive facilities in Springfield. An Amoskeag steamer called the *Constitution*, whose name later changed to *Henry Gray*, soon replaced the *Ocean*. In 1891, the city purchased the *Henry Gray* from the railroad company and used it for over 300 mutual aid calls to nearby towns and cities. On many of these mutual aid calls, the engine was transported to the fire on the special flatcar built to transport fire apparatus via the vast street railway network in the Springfield area. In case a large fire tied up the city fire apparatus, the city fire department could call upon the commandant of the United States arsenal for the use of its fire apparatus that was stationed at the armory.

In 1861, a hand engine called the *Champion* was purchased and located at Indian Orchard. It was a Hunneman machine that was very popular with New England fire departments because they were built in Massachusetts. The first steam pumper in Springfield was purchased in July 1862 and was called the *Monitor*. This engine was traded to LaFrance Engine Company toward the purchase of a new steam pumper in 1896. Also in 1896, the United States government purchased a new steamer called the *Union*, which was stationed at the armory. This engine later was transferred to the Frankfort Arsenal in Philadelphia and was replaced by an Amoskeag engine that was manned by U.S. Army soldiers. In December 1862, a new Amoskeag steamer was put in service for Engine Company No. 4. In March of the following year, the *Cataract* engine company was disbanded. On September 1, 1865, the city purchased another steamer called the *L. O. Hanson*, which served the city until it was sold to Stafford Springs, Connecticut, in 1877. The two hand engines, *Niagara* and *Cataract*, were sold to Red Wing, Minnesota, in 1865. In 1867, a new independent fire company called the Alert was organized. This company had its engine house on Alert Street near Worthington Street. In 1868, the Pynchon Street station was opened. This building served as fire headquarters until the Dwight Street station was built in 1911.

The fire department changed from volunteer to paid in 1872. The volunteers formed the Springfield Veteran Fireman's Association, a social and fraternal organization. From 1872, the fire department made great strides especially in the area of apparatus. On March 22, 1871, it purchased a new Junkett steam pumper. In December of the same year, a large Amoskeag steamer was acquired. The largest Amoskeag pumper the department purchased was an extra first size (today this would be a 1,250-gallon-per-minute pumper) in 1894. It was so large it had to be drawn by three horses. The first hook and ladder truck was purchased in 1871 and was stationed at Sanford Street. A new aerial ladder was put in service at Pynchon Street in June 1888. In February 1896, a new Seagrave cities service ladder truck was purchased and stationed in the Mill Street firehouse. The first chemical engine was purchased in 1894; it was a Holloway chemical and hose wagon assigned to Chemical Company A. The department would eventually operate five chemical companies in the horse-drawn era. By 1897, the department was operating out of 10 stations in the city.

In 1906, Springfield entered the motorized age with the introduction of a Knox squad truck. It carried a "flying squad" of eight firefighters and a battalion chief who responded to all box alarms in the city. By 1909, the city had five Knox trucks in service, a combination hose and chemical wagon, two flying squads, and two cars for the chief and his assistant. In 1913, the first motorized pumper was received, and by 1916, the four ladder trucks were all replaced by motorized apparatus. The department was fully motorized in 1916 and was the first major city fire department in the United States to do so. In the early years, Knox and Seagrave were the major suppliers of apparatus to the city but, over time, Mack and Maxim apparatus began to replace the older equipment when it was due for replacement. The department eventually grew to operate out of 14 stations, however, by 2006, population changes and increased labor costs left the department with only nine stations. The Springfield Fire Department is headed by fire commissioner Gary Cassanelli and has 252 career firefighters and staff represented by the International Association of Firefighters Local 648. In 2005, the department responded to 10,793 calls.

One

THE VOLUNTEER ERA

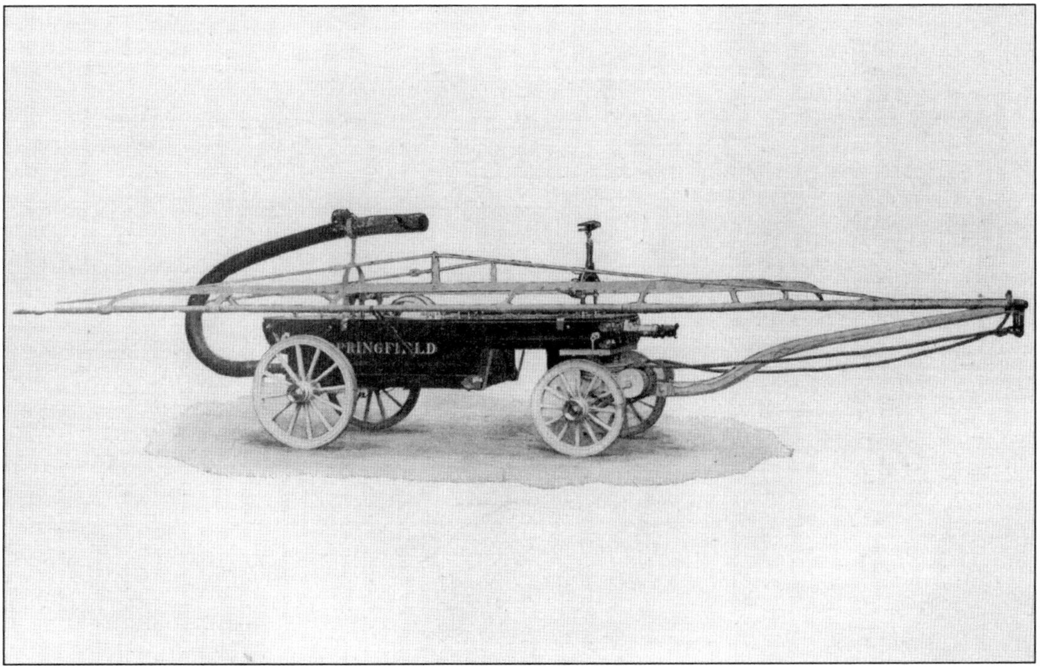

Pictured here is one of the early hand-drawn engines used during the volunteer era in Springfield. The curved tube is the suction hose that would be put into a cistern or pond to allow the engine to draw water into the piston pump. These were called "squirrel tails" by the early firefighters. The long poles on the side are the handles or brakes that the firemen moved up and down to operate the piston pump. Piston pumps have been around since the time of the Greeks and Romans, and the design is over 2,000 years old. (Courtesy Raymond Pond.)

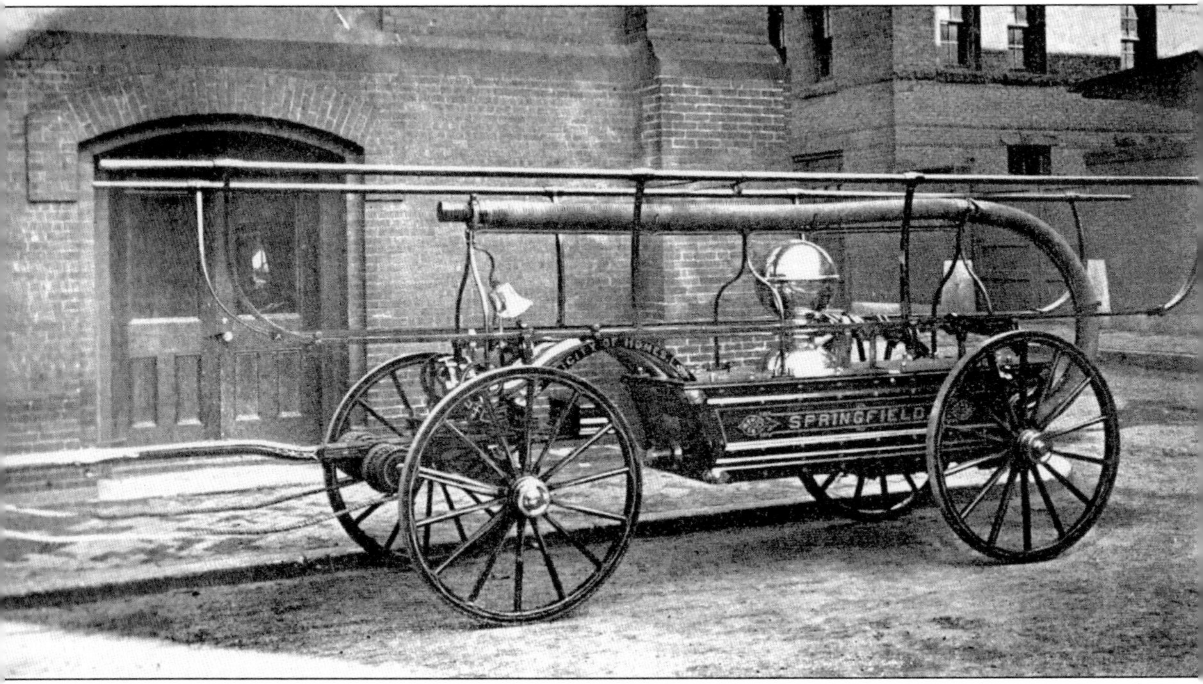

City of Homes was built by Button and Blake in 1859 for Red Wing, Minnesota, and was purchased secondhand by Springfield. This engine is now on display at the Connecticut Valley Historical Society in Springfield. (Courtesy Raymond Pond.)

Chiefs of the Springfield Fire Department
Volunteer Era

Elijah Blade	1830-1844
Cicero Simons	1845-1846
George Dwight	1848
James D. Brewer	1852-1853
George Ensworth	1854
Levi W. Park	1855
George Dwight	1856-1859
Hosea C. Lombard	1860
Levi O. Hanson	1861
Joseph Marsh	1862
O.S. Taylor	1862
Lucius H. Powers	1865-1868
W.W. Day	1868
Lucius H. Powers	1869
Abner P. Leshure	1870

Shown is a list of chiefs who served during the volunteer era of the Springfield Fire Department.

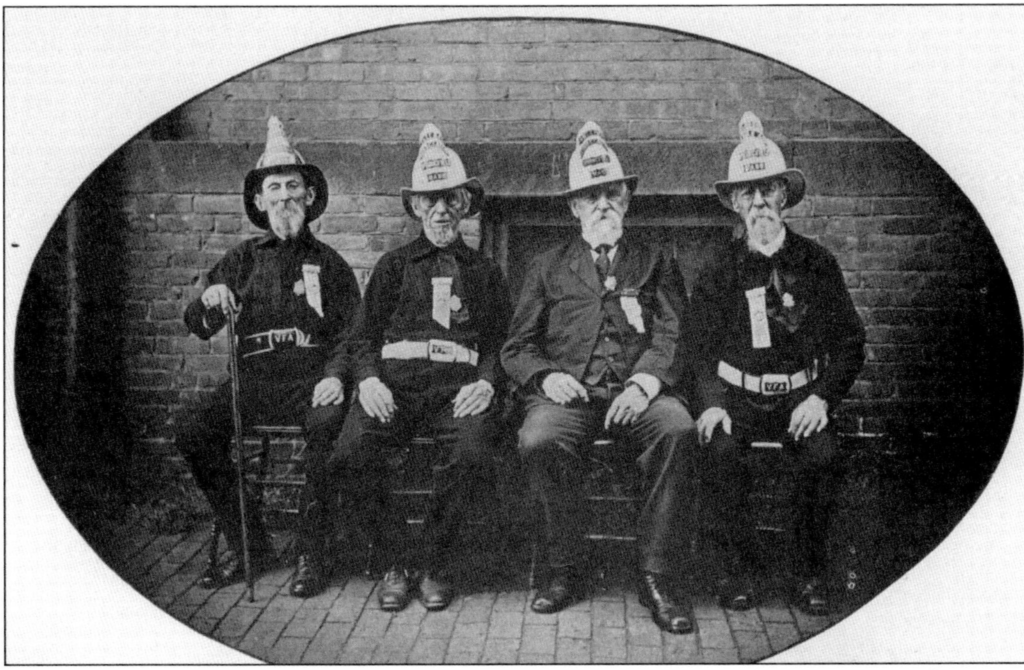

Former members of the volunteer fire department are shown in this 1898 view. (Courtesy Connecticut Fire Museum archives.)

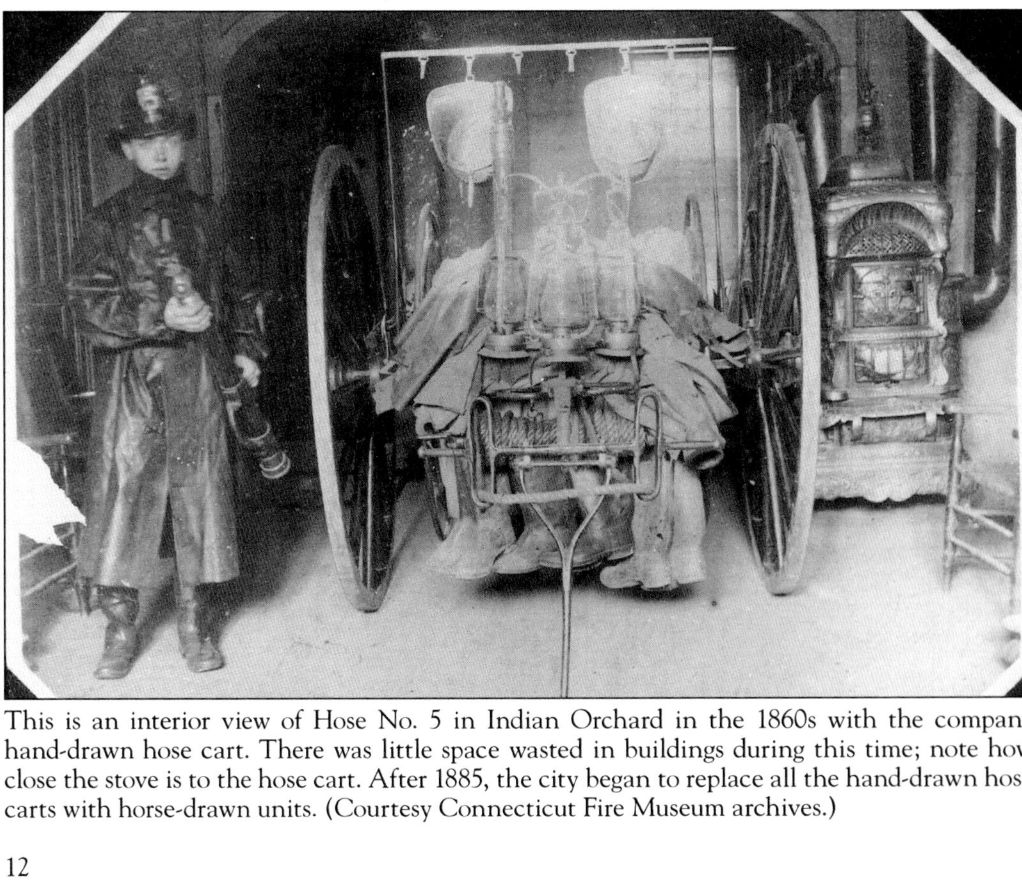

This is an interior view of Hose No. 5 in Indian Orchard in the 1860s with the company hand-drawn hose cart. There was little space wasted in buildings during this time; note how close the stove is to the hose cart. After 1885, the city began to replace all the hand-drawn hose carts with horse-drawn units. (Courtesy Connecticut Fire Museum archives.)

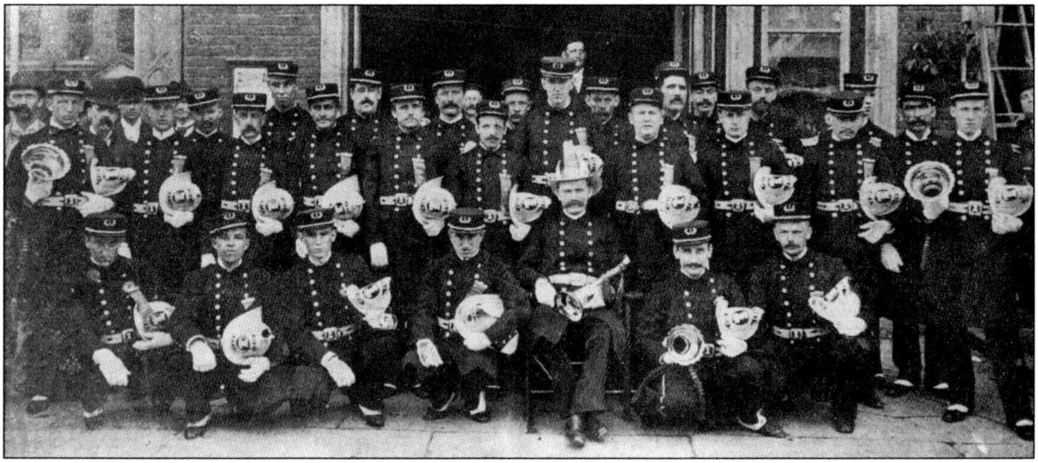

This 1890 picture shows the members of the Springfield Veteran Fireman's Association. (Courtesy Connecticut Fire Museum archives.)

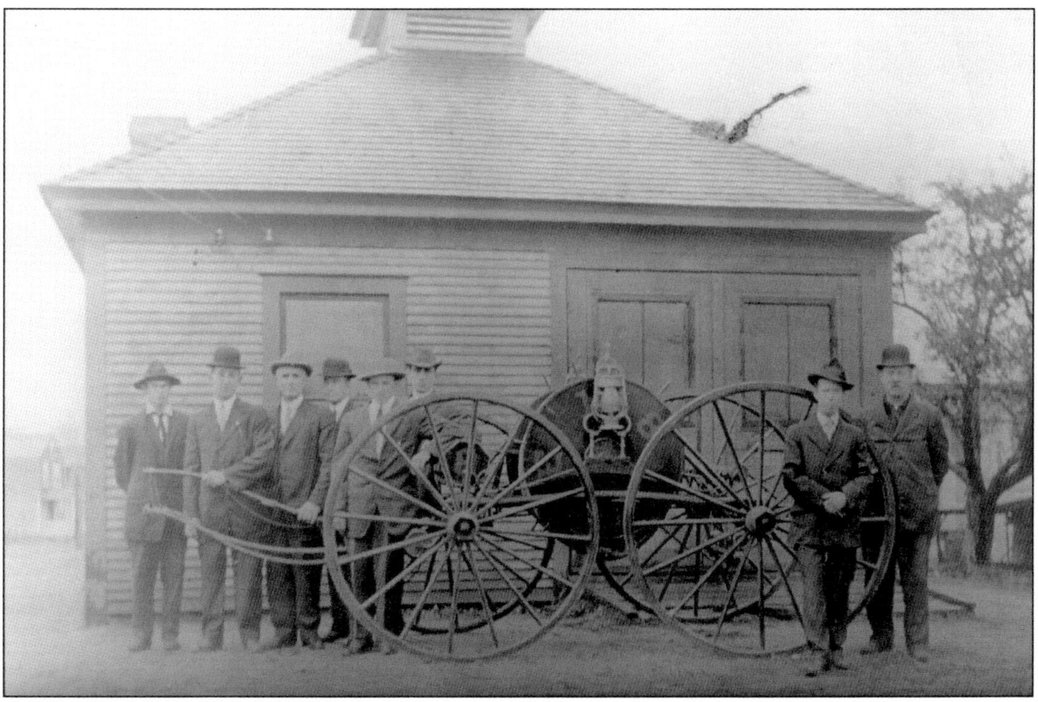

This is a view of the Pine Point Volunteer Fire Company firehouse that was located on Jasper Street near Boston Road. Today Engine Company No. 16, located on Massreco Street, covers this area. (Courtesy Raymond Pond.)

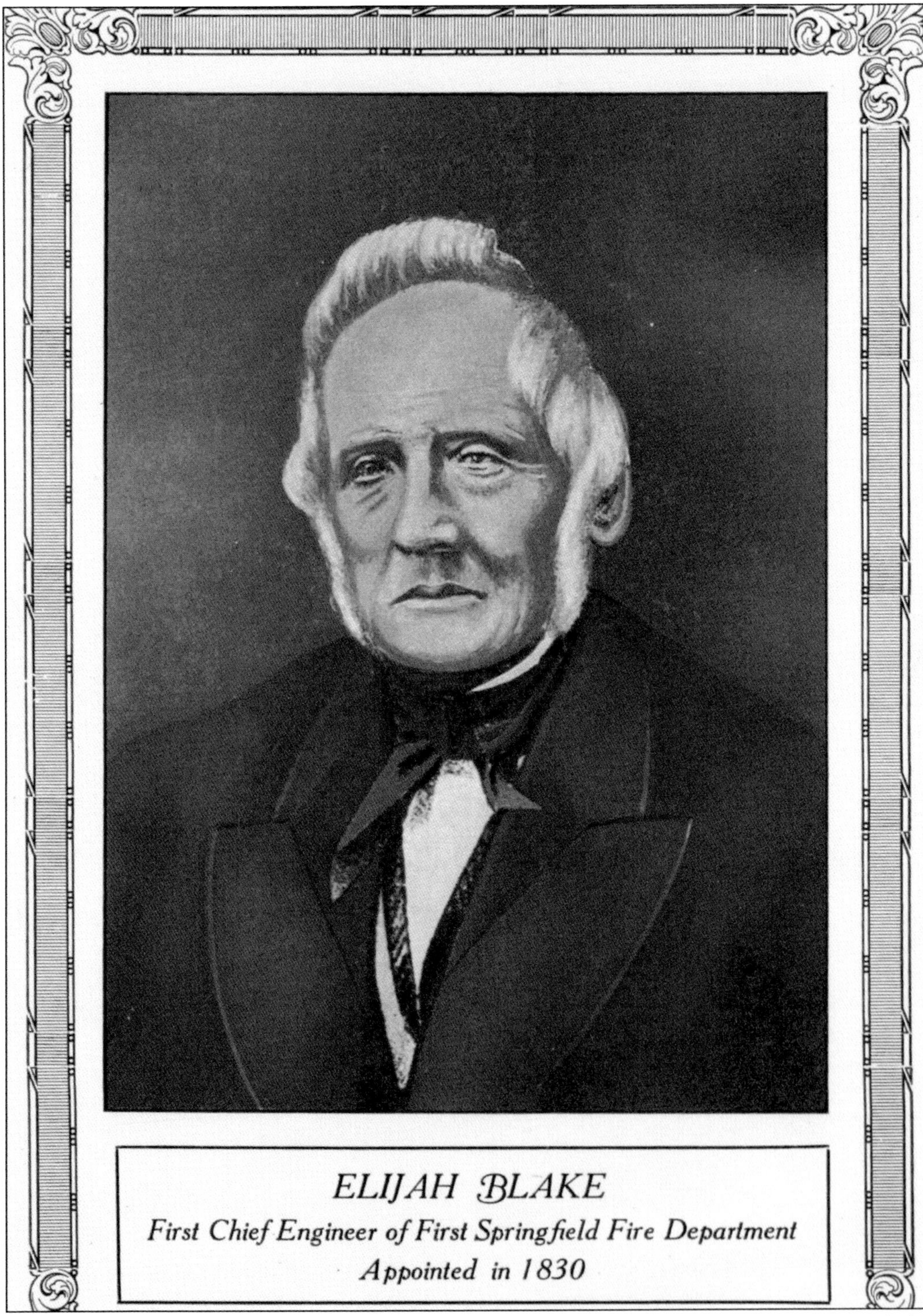

Elijah Blake was the first chief engineer of the Springfield Fire Department. He joined the department in 1809 and served as chief from 1830 to 1844. When he retired in 1844, he was 96 years old. (Courtesy Raymond Pond.)

Two

The Horse-Drawn Era

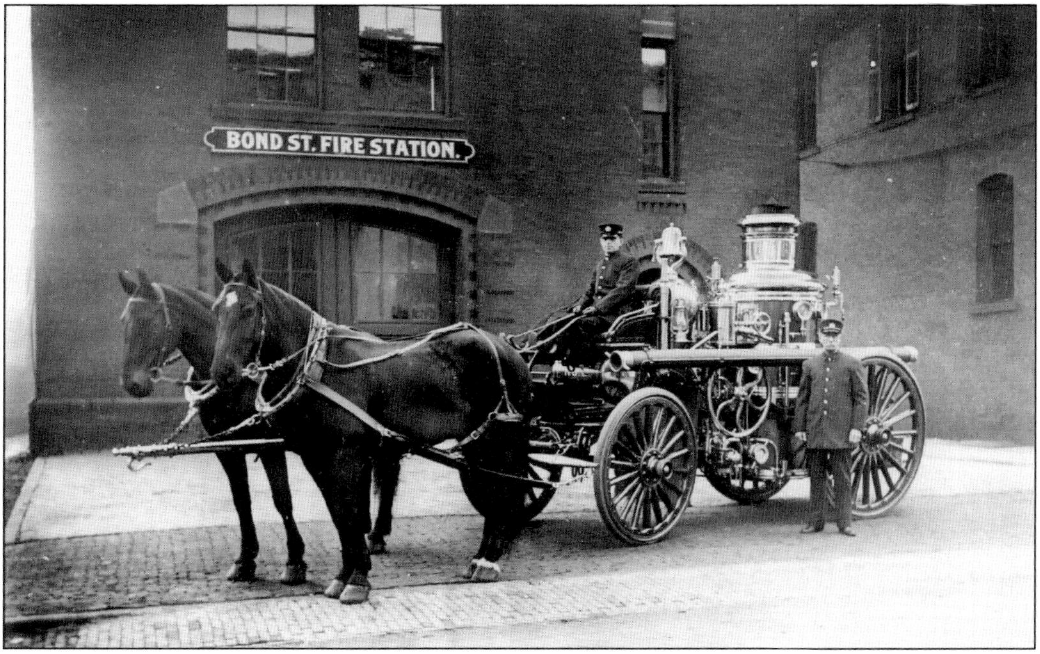

This view shows the steamer of Engine Company No. 2 that operated out of the Bond Street fire station. The fire station opened in 1888 and closed in 1910, when the company was relocated to North Street (later renamed Dwight Street). Steam fire engine manufacturing was popular in the United States from 1853 to 1913. The last one was built in 1917 for the Pittsburgh Fire Department. (Courtesy Noah Pooler family archives.)

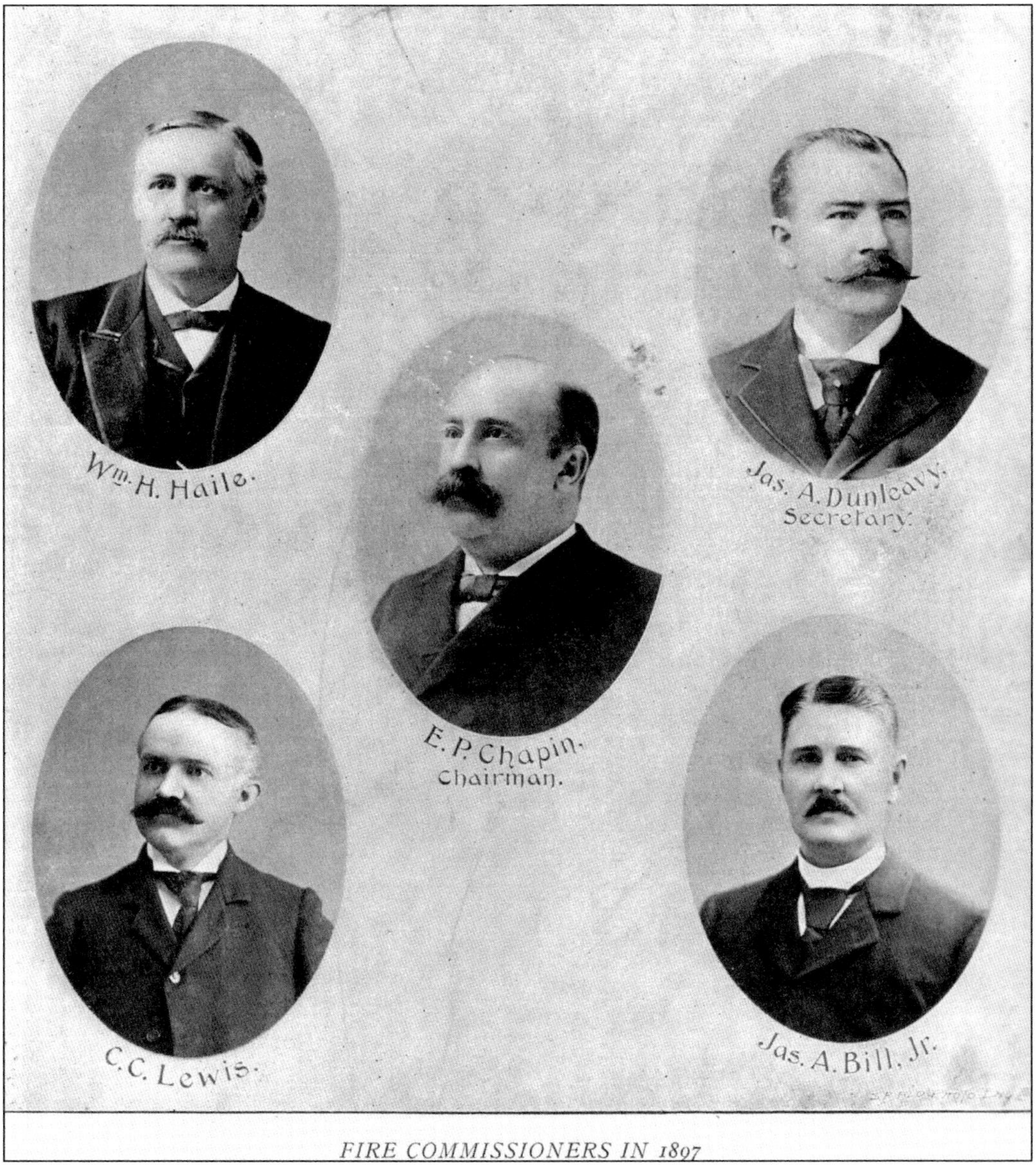

Shown are the members of the Springfield Fire Department Board of Fire Commissioners in 1897. (Courtesy Raymond Pond.)

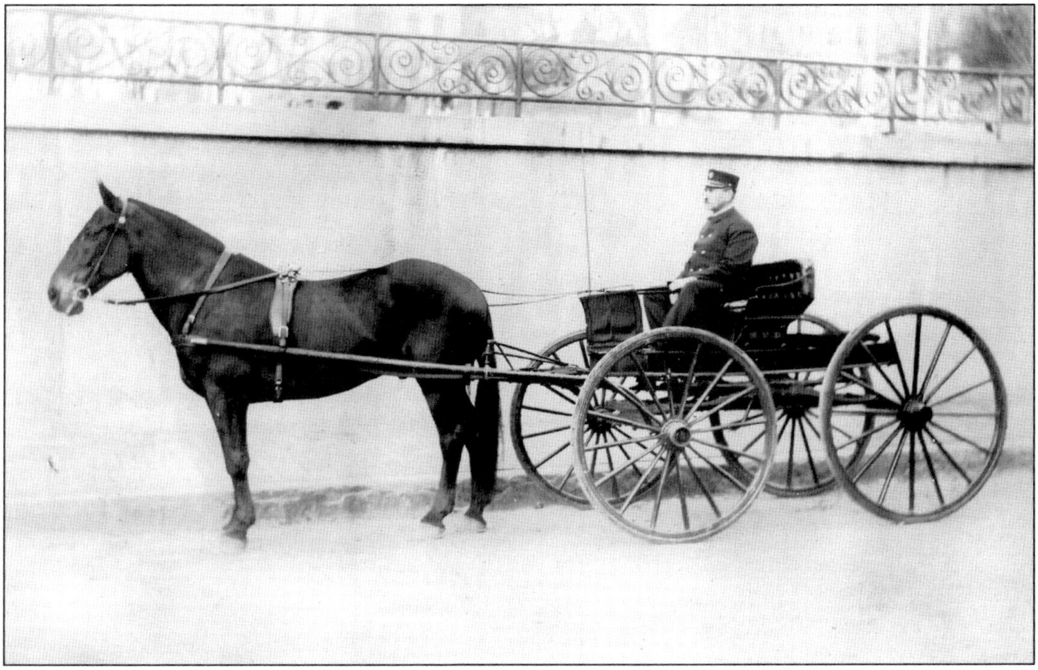

The chief and his deputies responded to fires by buggy until the department purchased some Knox automobiles in 1906. (Courtesy Raymond Pond.)

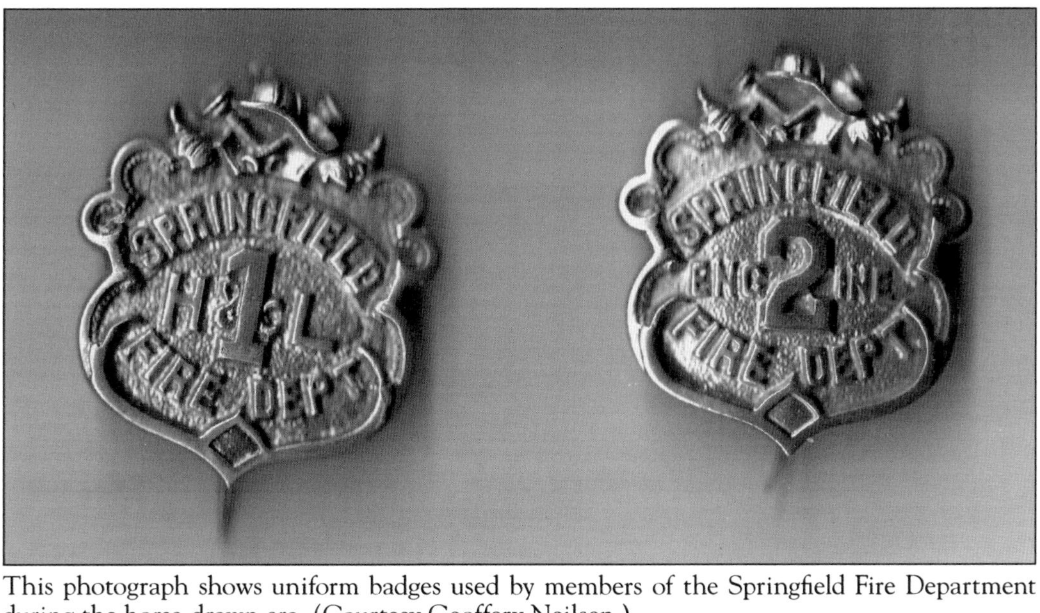

This photograph shows uniform badges used by members of the Springfield Fire Department during the horse-drawn era. (Courtesy Geoffery Neilsen.)

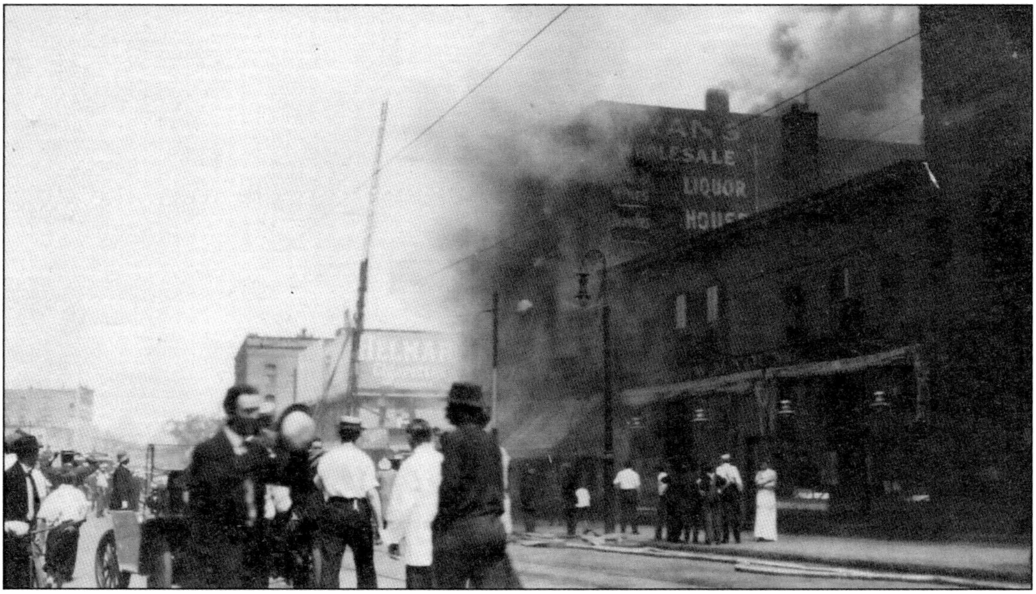

In this view of the Regan Building fire that occurred in 1915, the horse-drawn hose wagon has laid hose lines to the fire and the men are using ground ladders to attack the blaze. (Courtesy Noah Pooler family archives.)

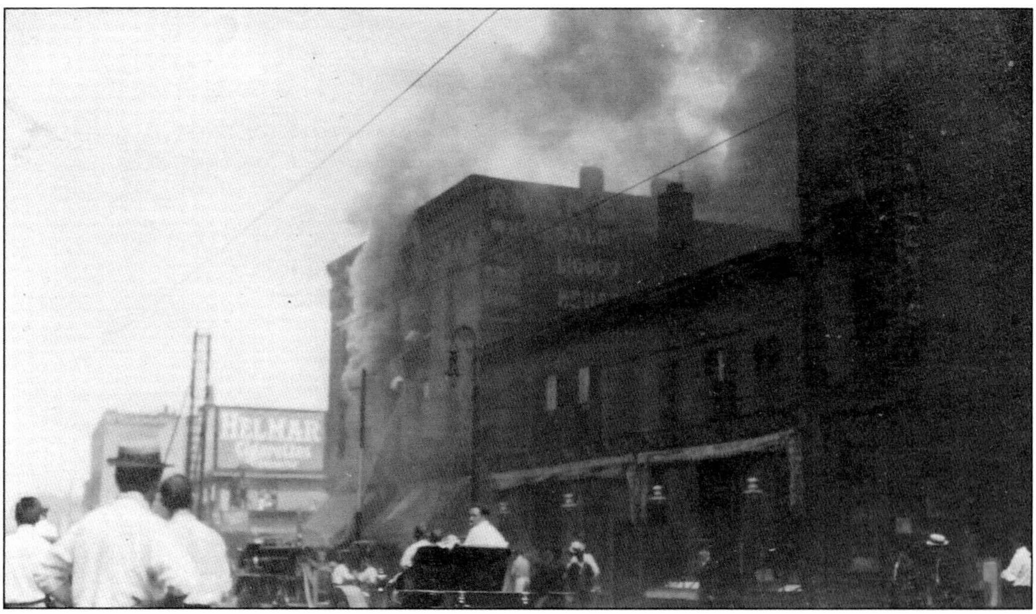

In this view of the fire, the ladder company men are raising their 50-foot extension ladder. It took at least four firemen to raise this ladder, which weighted over 500 pounds. (Courtesy Noah Pooler family archives.)

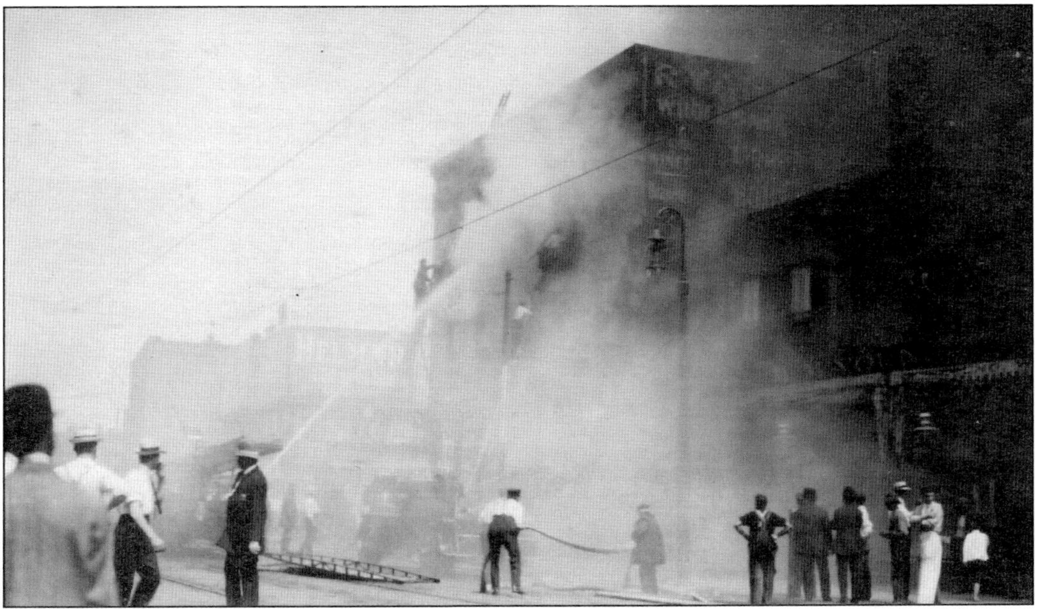

In the era of the streetcar, the overhead wires that supplied electricity to the streetcars were charged with 600 volts of direct current. This made for a dangerous situation if the firefighters were not careful where they directed their water streams. (Courtesy Noah Pooler family archives.)

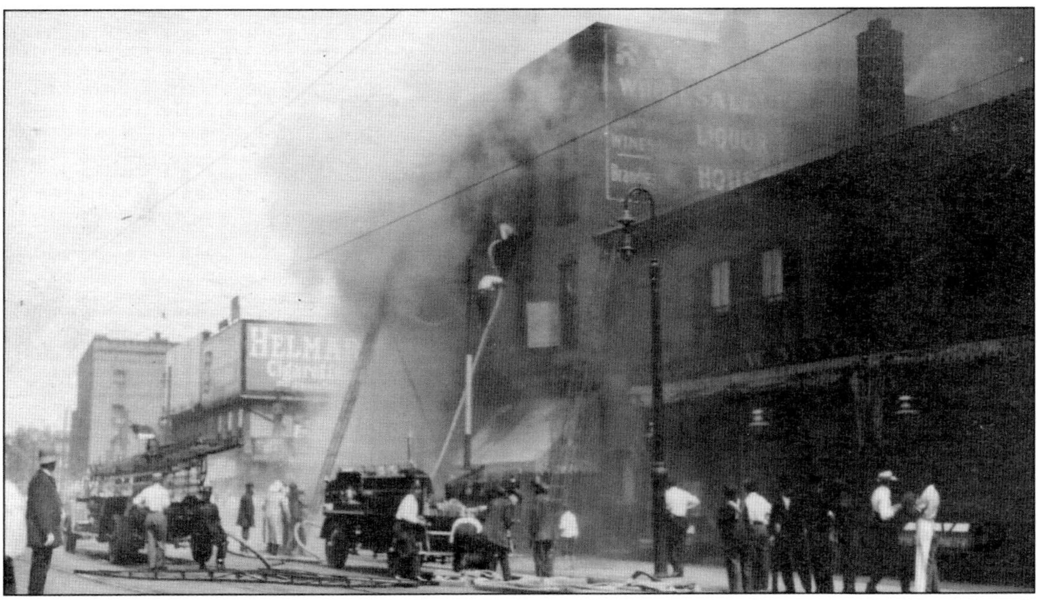

In this photograph, the firefighters have three streams of water on the Regan Building fire. A short time after the photograph was taken, the fire was under control. This was before the era of air-breathing equipment, which would not be available to the fire service for another 60 years. Many firemen of this era died from breathing in smoke and dangerous fumes from the burning building and its contents. (Courtesy Noah Pooler family archives.)

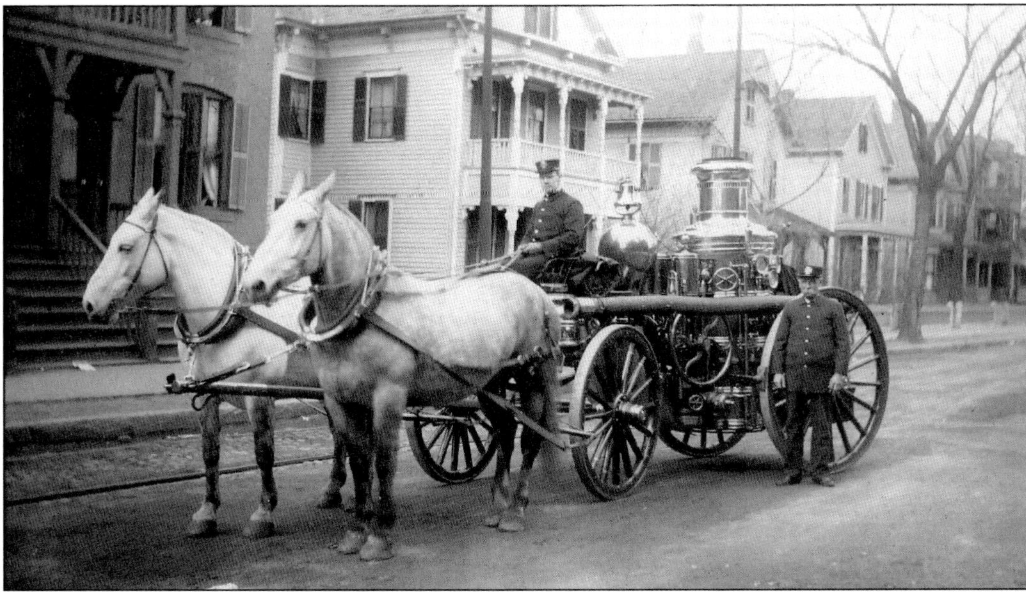

The steamer of Engine Company No. 2, with its driver and engineer, is shown in this c. 1905 view at the Bond Street fire station. The pear-shaped dome behind the driver is an air chamber that evens out the pulsations of water flowing from the pistons driving the pump. This allows the water to be discharged in a steady stream rather than in intermittent spurts. Today the Springfield social security office occupies the site behind the engine. (Courtesy Noah Pooler family archives.)

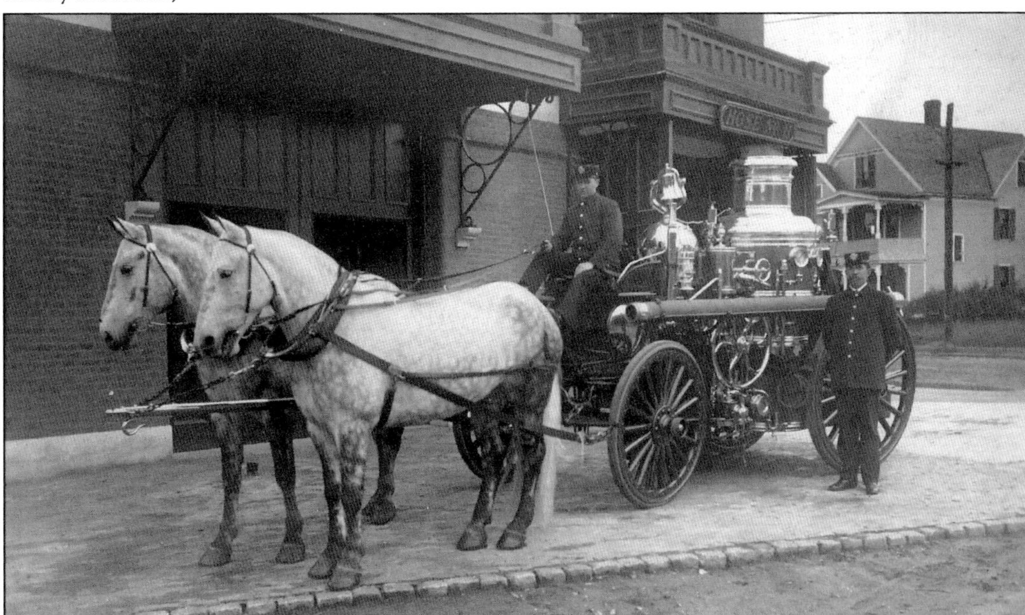

This 1910 view shows Engine Company No. 3 in front of the Oakland Street firehouse. Horse-drawn apparatus were the mainstay of the fire service in 1910. The steam pumper had reached the pinnacle of its development, and skeptical fire chiefs viewed the idea of gasoline-powered vehicles replacing horse-drawn apparatus as little more than a passing fad. The Oakland Street firehouse was still in service as a fire station in 2006. (Courtesy Noah Pooler family archives.)

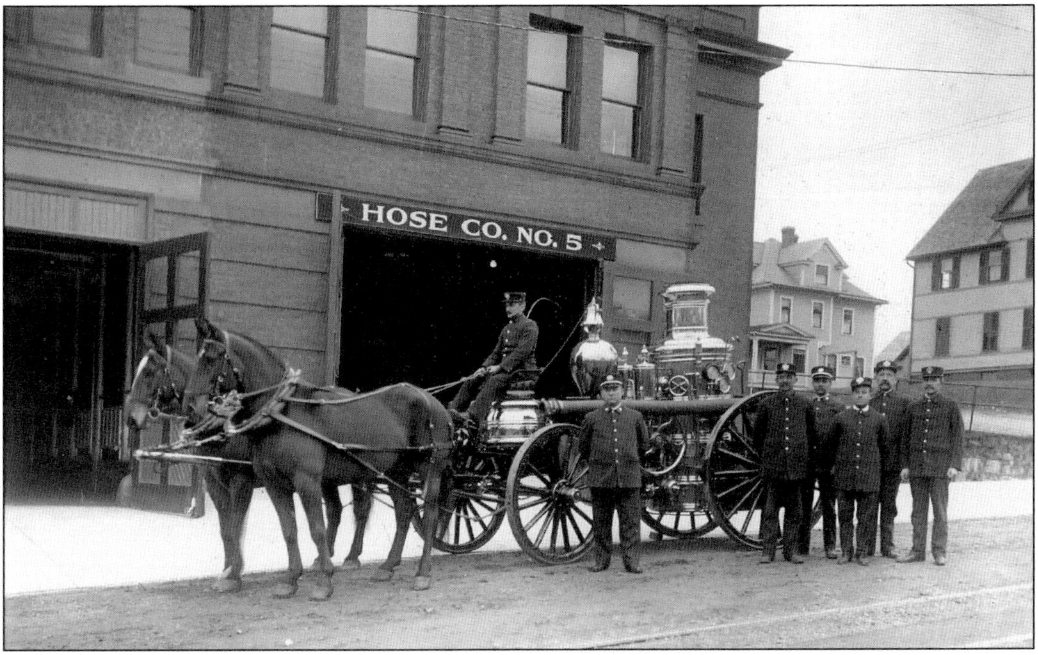

The Indian Orchard fire station with a steamer from Engine Company No. 5 is pictured here around 1910. These engines had the boiler sheathed in nickel to help transform the engine from a plain piece of machinery into a dazzling showpiece. Springfield spared no expense when it purchased an engine. (Courtesy Noah Pooler family archives.)

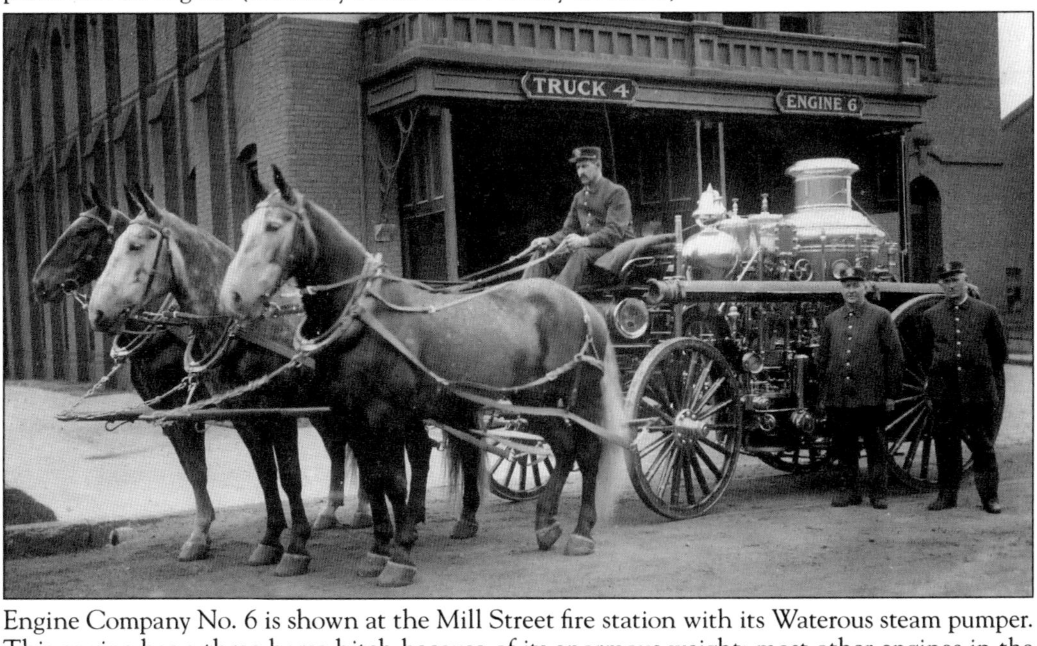

Engine Company No. 6 is shown at the Mill Street fire station with its Waterous steam pumper. This engine has a three-horse hitch because of its enormous weight; most other engines in the Springfield department only had two-horse hitches. The stalls, feed storage, and coal storage occupied the rear of the first floor. Over this was a large hayloft. The firefighters had their bunk room and day room on the second floor and used the familiar brass pole to get to the first floor during an alarm. (Courtesy Noah Pooler family archives.)

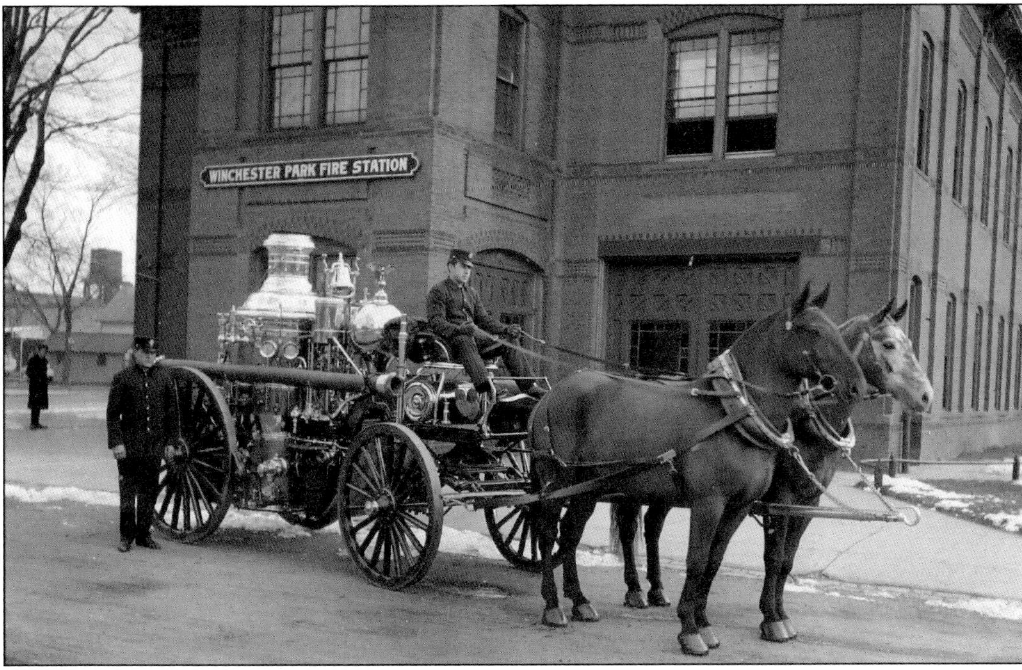

The Winchester Park fire station with the 1896 American LaFrance steam pumper and driver Percy Osborn from Engine Company No. 8 is shown in this view. Routine firehouse duties included daily care and exercise of the horses. (Courtesy Noah Pooler family archives.)

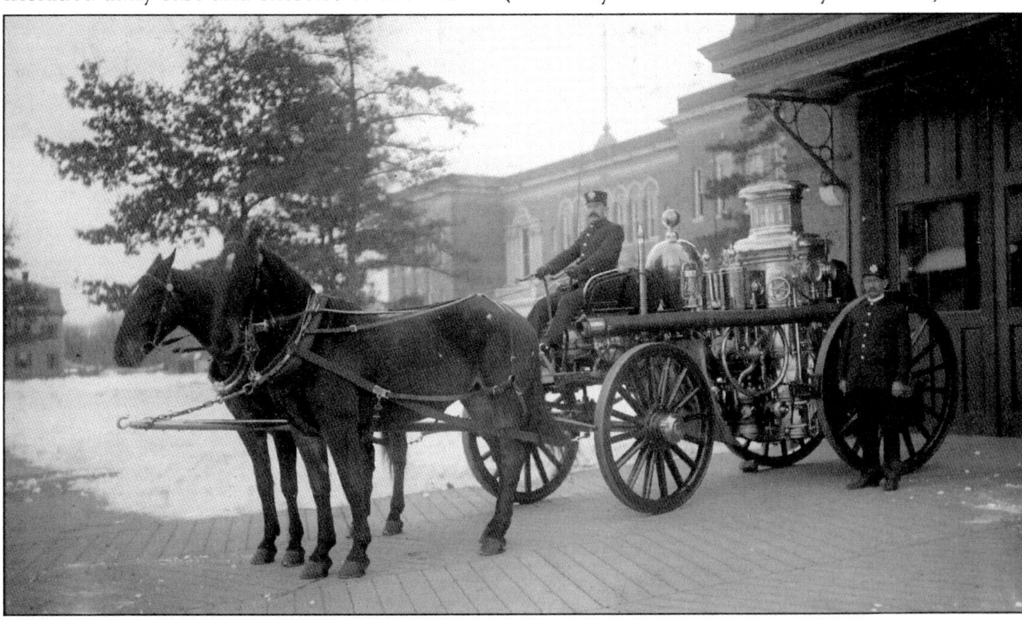

The steamer from Engine Company No. 9 is shown at the Armory Street fire station around 1910. The station was opened in 1902 and closed in 2002. The Armory Street school is shown in the background. When in the firehouse, the steamer was connected to a small stationary boiler that kept the water in the steamer's boiler hot. After the company left the fire station, the engineer on the rear platform lit a combination of wood and coal in the firebox so that steam would be up when they reached the fire scene. (Courtesy Noah Pooler family archives.)

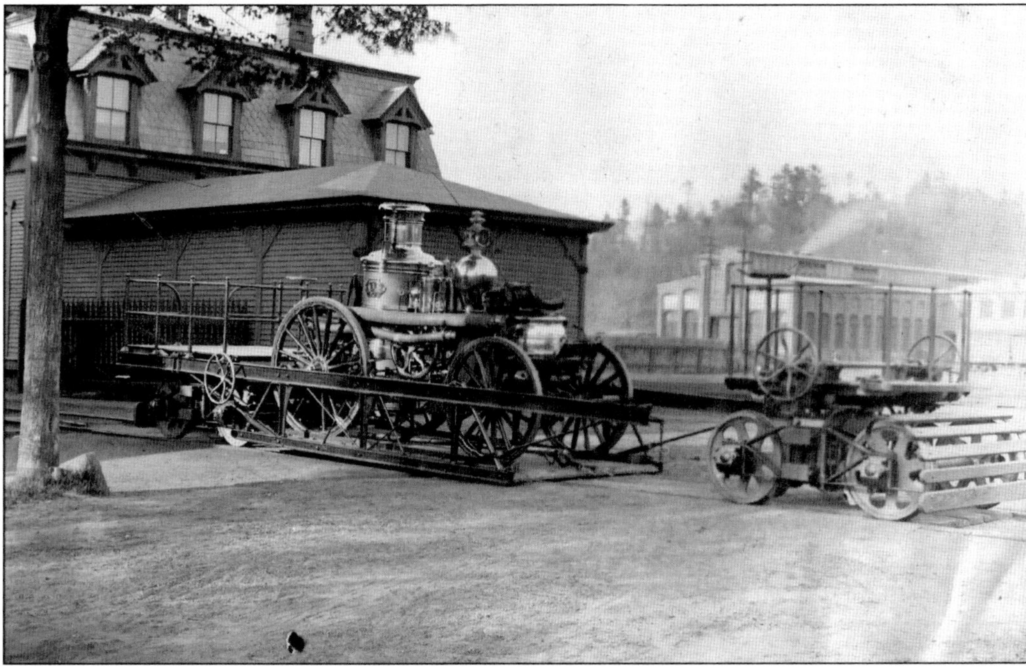

To provide horse-drawn steam fire engines quicker access to the outlying points, this vehicle was conceived in 1895 by Fire Commissioner Edward P. Chaplin and Chief W. J. Littlefield of the Springfield Fire Department. It was designed for the quick transport of horse-drawn steam fire engines over the street railway lines that were then beginning to spread out from Springfield. Built by the Wason Manufacturing Company in 1897, the car was 30 feet long, weighed seven tons, and was designed to be pulled by a streetcar. (Courtesy of Noah Pooler family archives.)

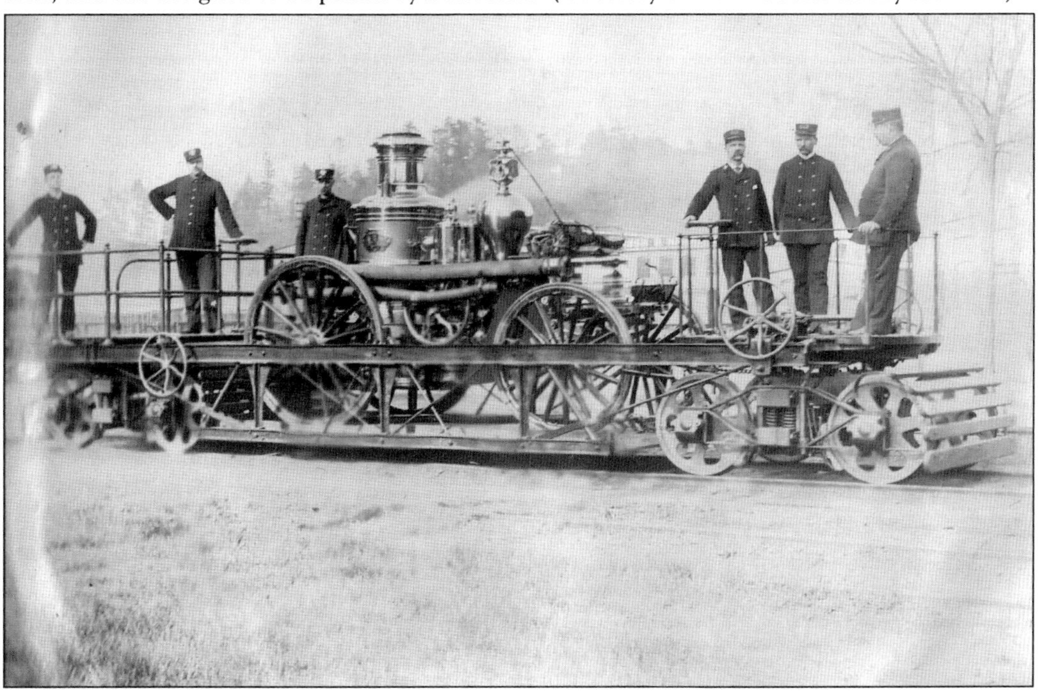

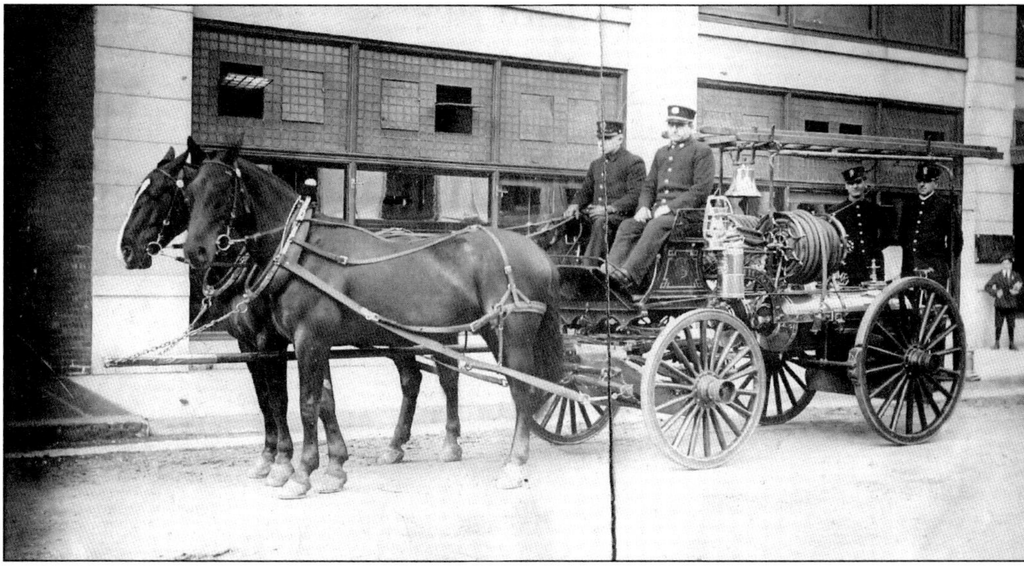

The 1894 Holloway chemical engine of Chemical Company B had two chemical cylinders so the operator could be recharging one while the other was being used. A two-cylinder unit could work indefinitely as long as the operator had a supply of water, soda, and acid. Chemical Company B operated out of the Pynchon Street fire station. Chemical engines originated in France to fight small fires. Upon arrival at the fire, acid was added to a soda mixture, creating pressurized foam that consisted of water and carbon dioxide, with the mixture fed to the fire through a one-inch rubber hose. Soda acid and chemicals are no longer used by the fire service. (Courtesy Noah Pooler family archives.)

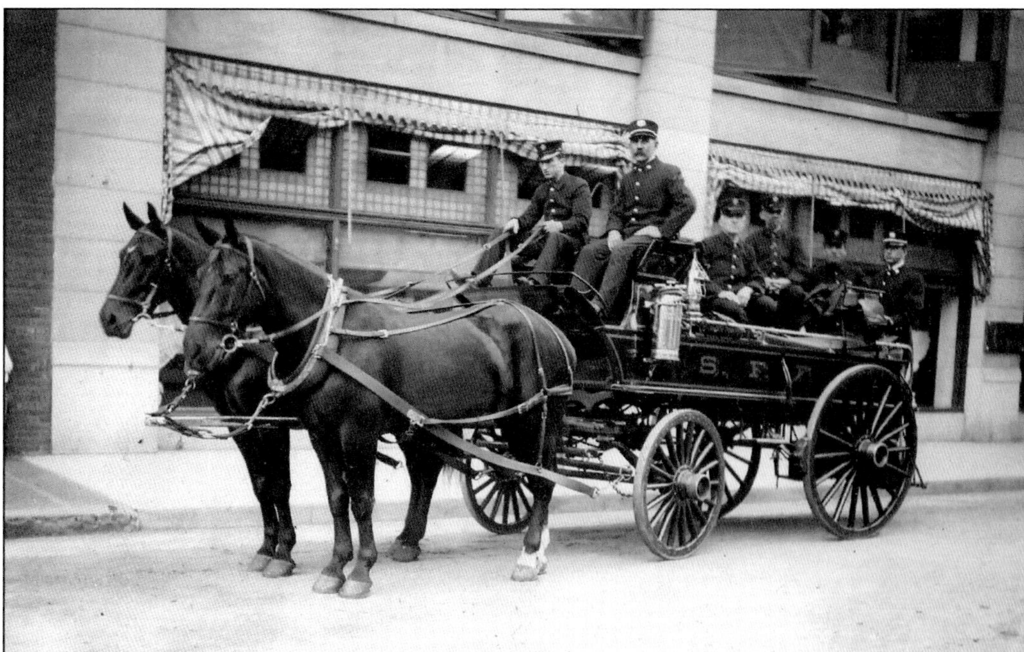

Hose Company No. 1 also operated out of the Pynchon Street firehouse using this two-horse hose wagon pictured around 1905. (Courtesy Noah Pooler family archives.)

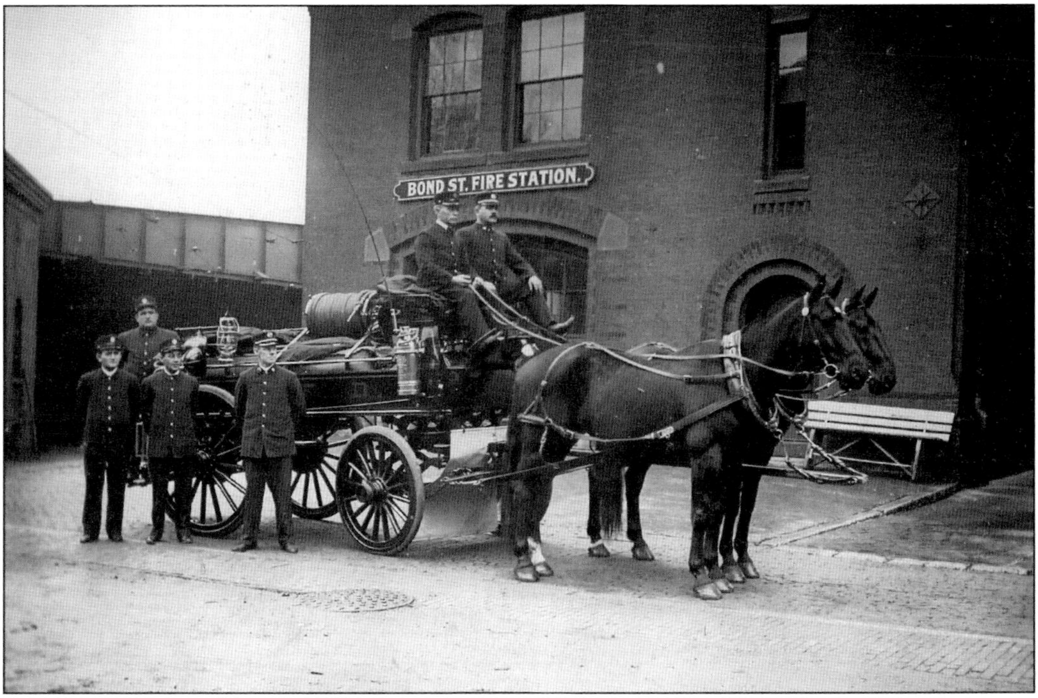

Hose Company No. 2 operated out of the Bond Street fire station along with Engine Company No. 2. As steamers did not carry any hose or equipment, a hose wagon always responded to the fire with the steamers. (Courtesy Noah Pooler family archives.)

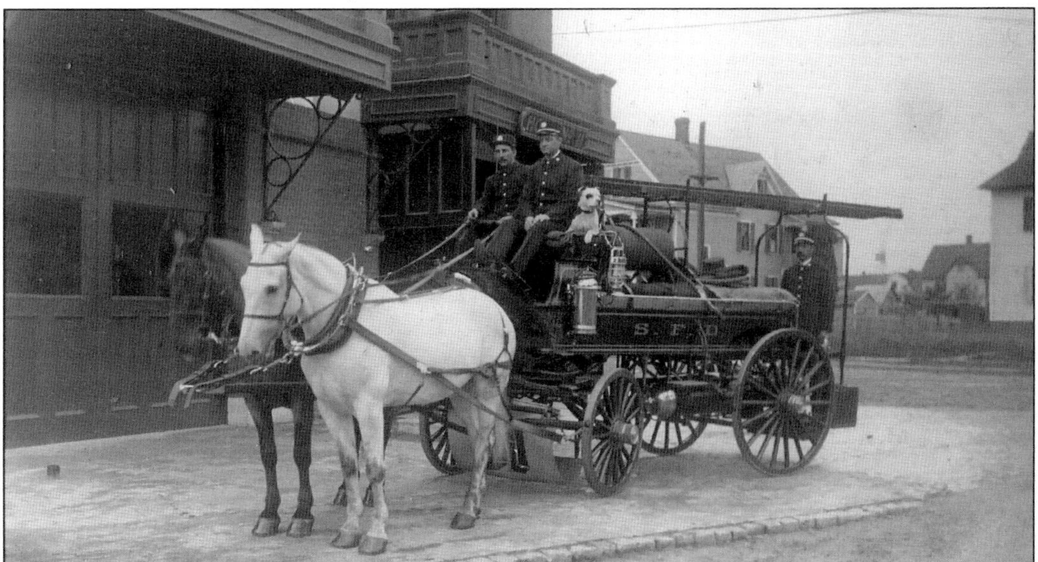

Engine Company No. 3 and Hose Company No. 3 both operated out of the Oakland Street fire station. The dog on the wagon was an important member of the company by providing security when the company was out on a call and helping to control the mouse and rat problem as many fire stations in this era were basically barns. The driver is unidentified, but the gentleman on the right was Fire Captain Boss. Captain Boss was considered a hero when he rescued three people at the Gilmore Opera House fire on April 5, 1897. (Courtesy Noah Pooler family archives.)

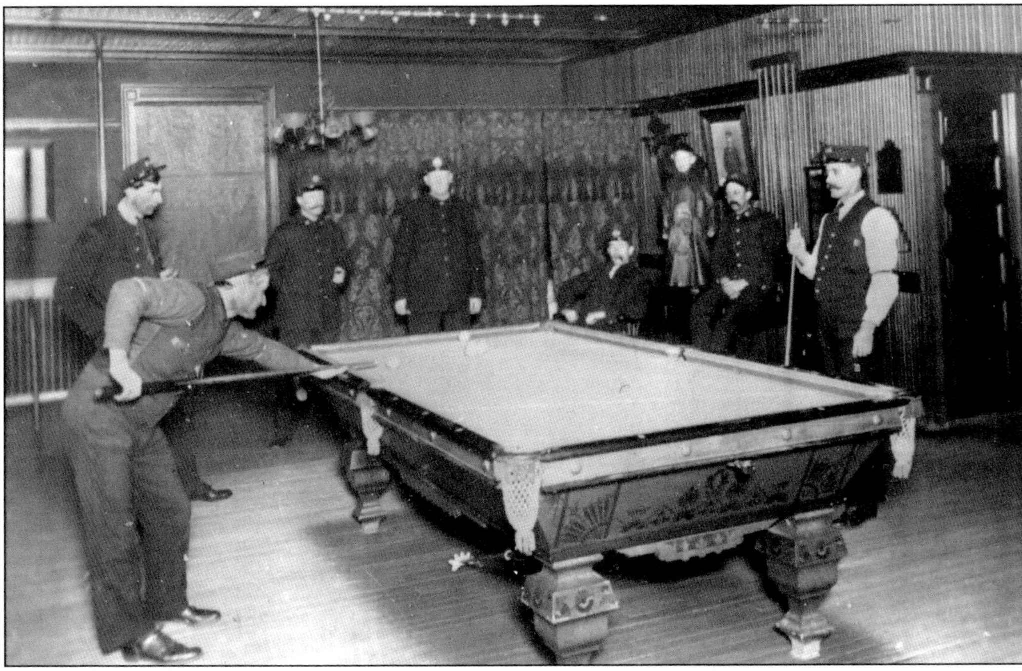

The day room on the second floor of Hose Company No. 4 on Walnut Street had this nice pool table to keep the firemen occupied between calls in the days before television or radio. Card games were also popular, especially pinochle. (Courtesy Noah Pooler family archives.)

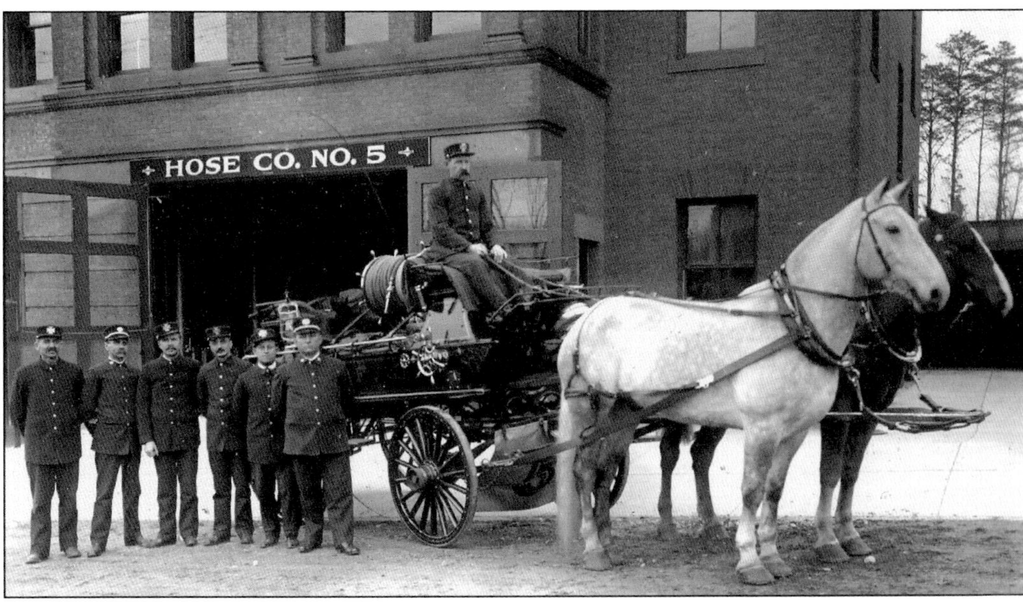

Indian Orchard housed a steam pumper and hose wagon along with the firefighters and four horses that pulled the apparatus. Horses were stabled in the rear, with living quarters on the second floor. Fire horses were not of any particular breed, but they had to be strong and fast. Each horse had a name and was given special care at all times. They were visited regularly by the department veterinarian and ate about 35 pounds of oats and hay per day along with assorted horse goodies like carrots and apples. (Courtesy Noah Pooler family archives.)

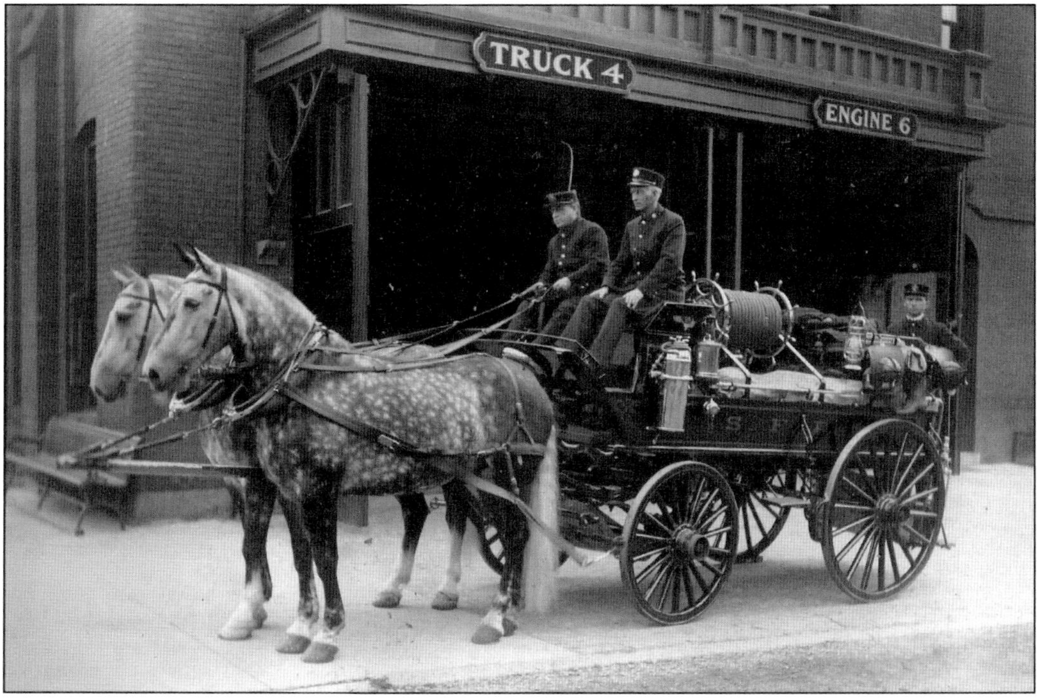

This view shows the chemical and hose wagon assigned to Engine Company No. 6 at the Mill Street firehouse. (Courtesy Noah Pooler family archives.)

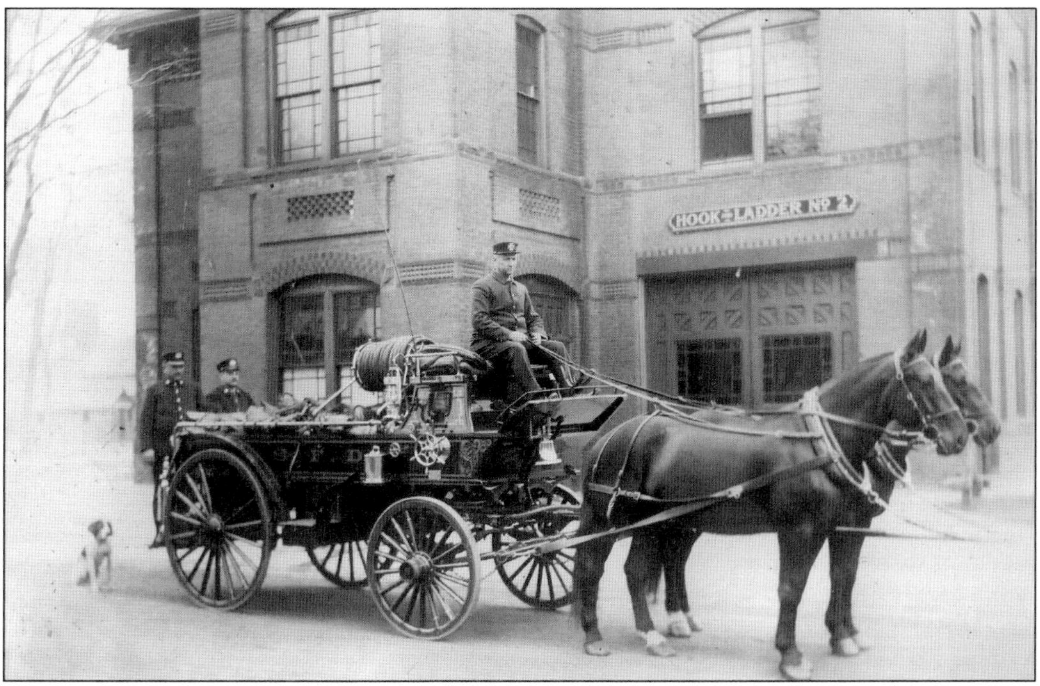

The 1891 Flynn and Kirby chemical and hose wagon of Engine Company No. 8 is shown outside the Winchester Park fire station. This station also housed Ladder Company No. 2. (Courtesy Noah Pooler family archives.)

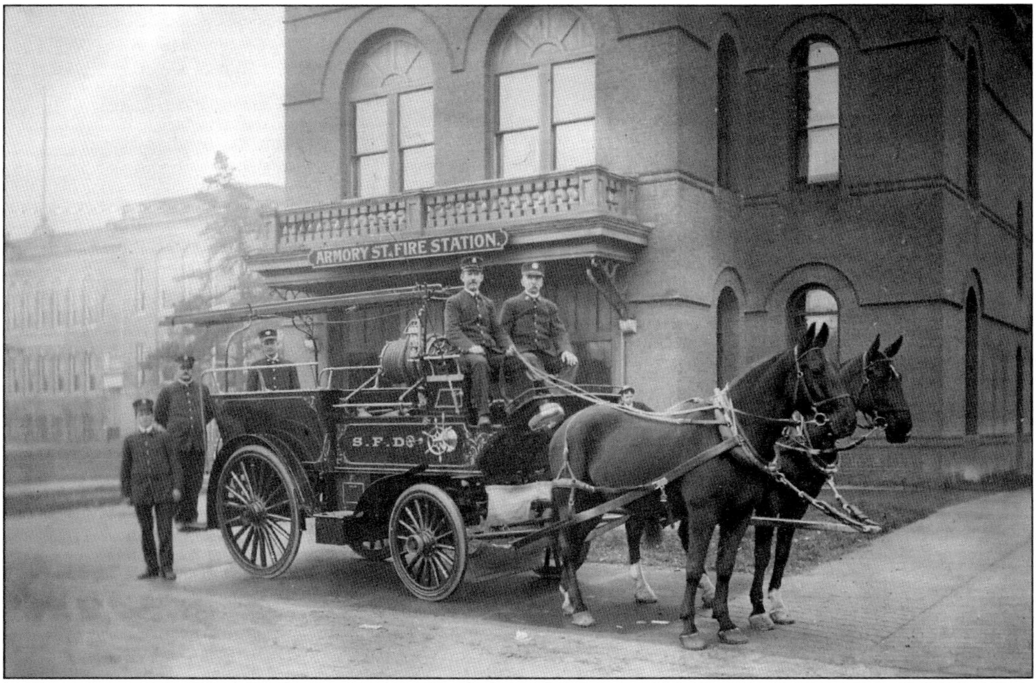

The 1893 Flynn and Kirby chemical and hose wagon from Hose Company No. 9 is shown in this view at the Armory Street fire station from about 1905. Hose Company No. 9 originally ran out of the North Main Street fire station. (Courtesy Noah Pooler family archives.)

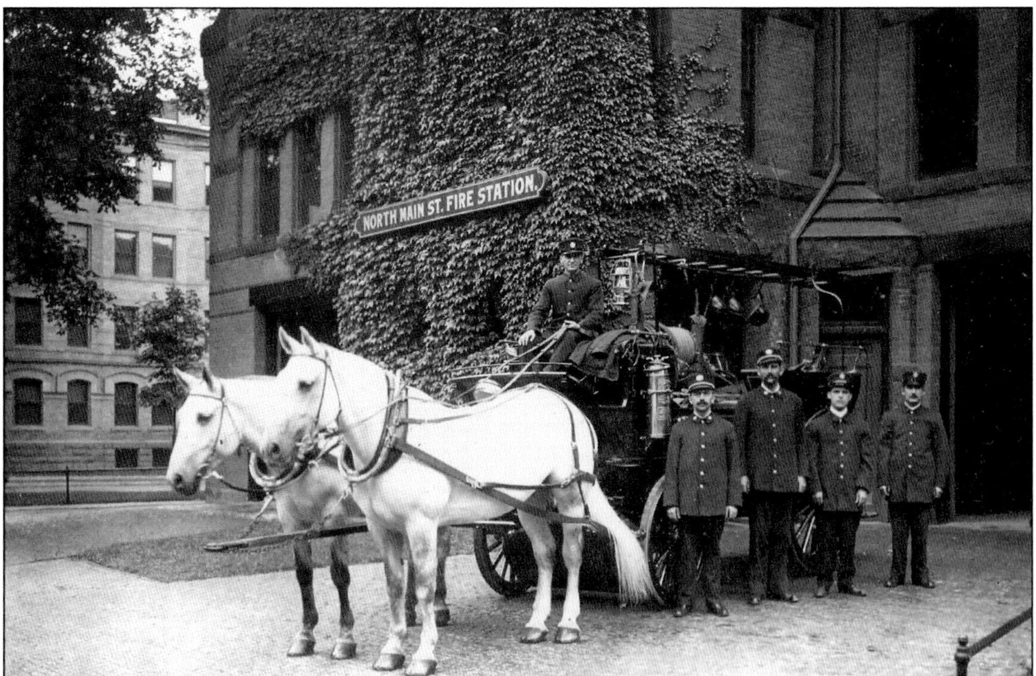

This 1910 view shows the hose wagon of Hose Company No. 10 at the North Main Street fire station that also housed Ladder Company No. 10. (Courtesy Noah Pooler family archives.)

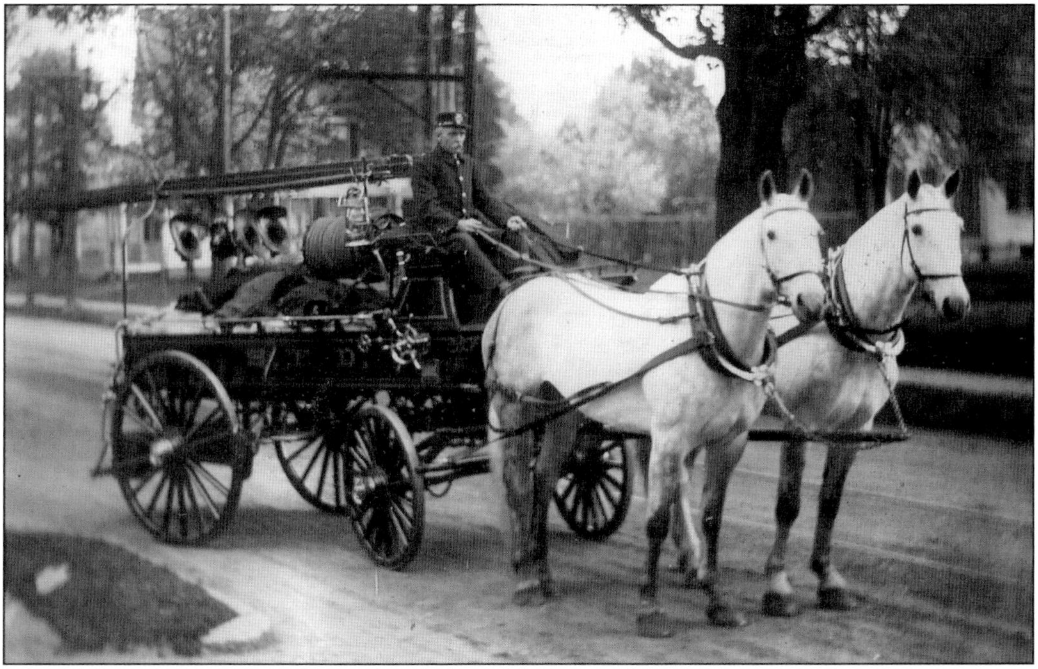

Hose Company No. 11 had this combination chemical and hose wagon pulled by two horses. Note that the street is not paved. (Courtesy Noah Pooler family archives.)

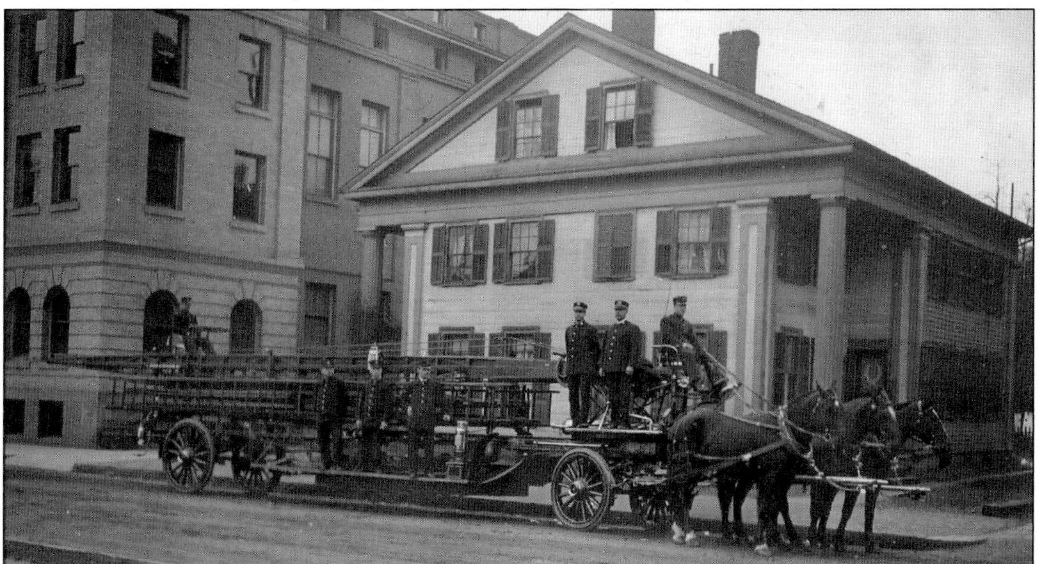

Ladder Company No. 1 operated this 1888 Champion aerial ladder truck with its three-horse hitch. Harnessing the three-horse hitch had special problems as the middle horse had to be trained to get into position first. If one of the other horses lined up first this well-trained middle horse pushed the other horse out of the way so he could get into the proper position. Firefighters who served in the horse-drawn era said that the horses usually lined up perfectly, but if something irritated the horses or something was out of place, they were nearly impossible to hitch up. This picture was taken in 1909 at the intersection of Main and Greenwich Streets. Lt. Noah Pooler is the second man to the left of the driver. (Courtesy Noah Pooler family archives.)

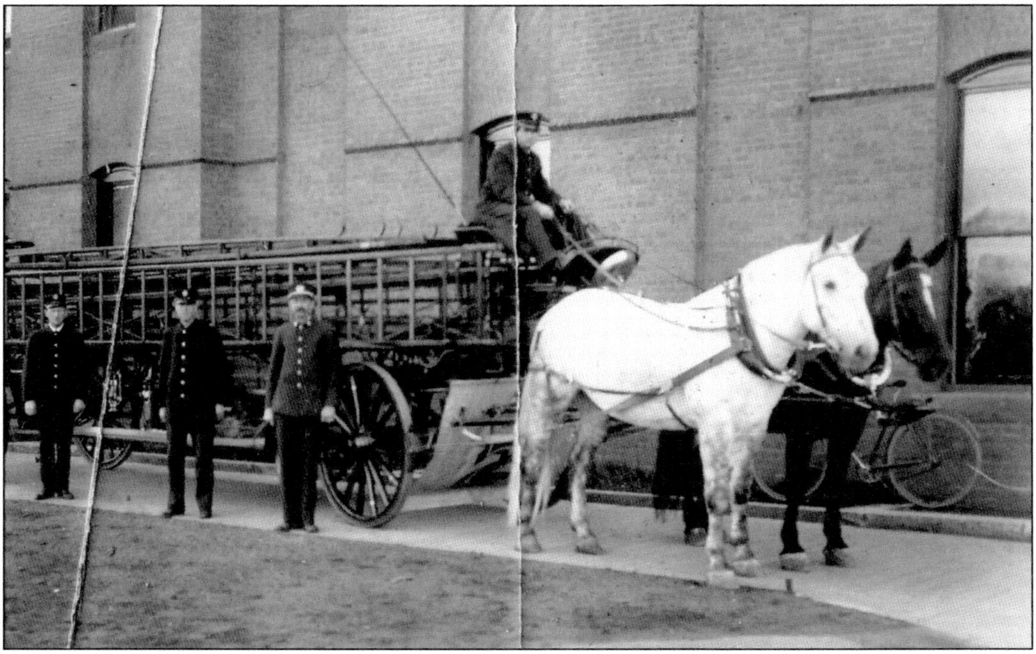

Ladder Company No. 3 used the city's service ladder truck that only carried ground ladders and operated out of the Oakland Street station with Engine Company No. 3. Notice the poop guard behind the front wheels that prevented horse manure from hitting the wagon. (Courtesy Noah Pooler family archives.)

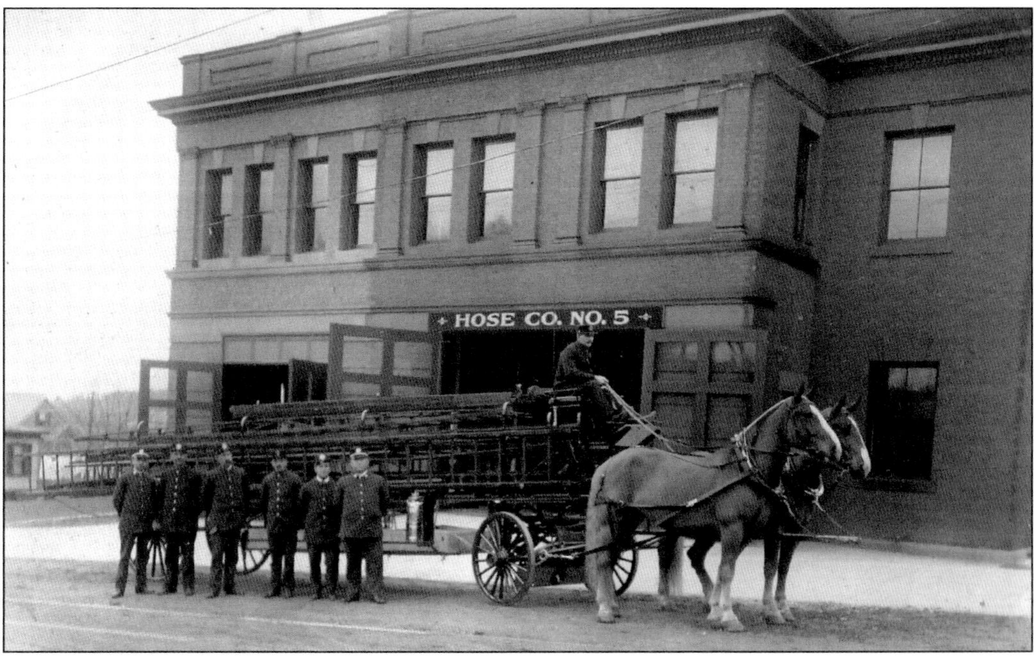

Ladder Company No. 5 shared the Indian Orchard fire station with Engine Company No. 5. By 1910, when this picture was taken, the city's service ladder trucks were already obsolete. Ladder companies carried a vast selection of tools, ropes, and other equipment needed for forced entry and rescue at fire scenes. (Courtesy Noah Pooler family archives.)

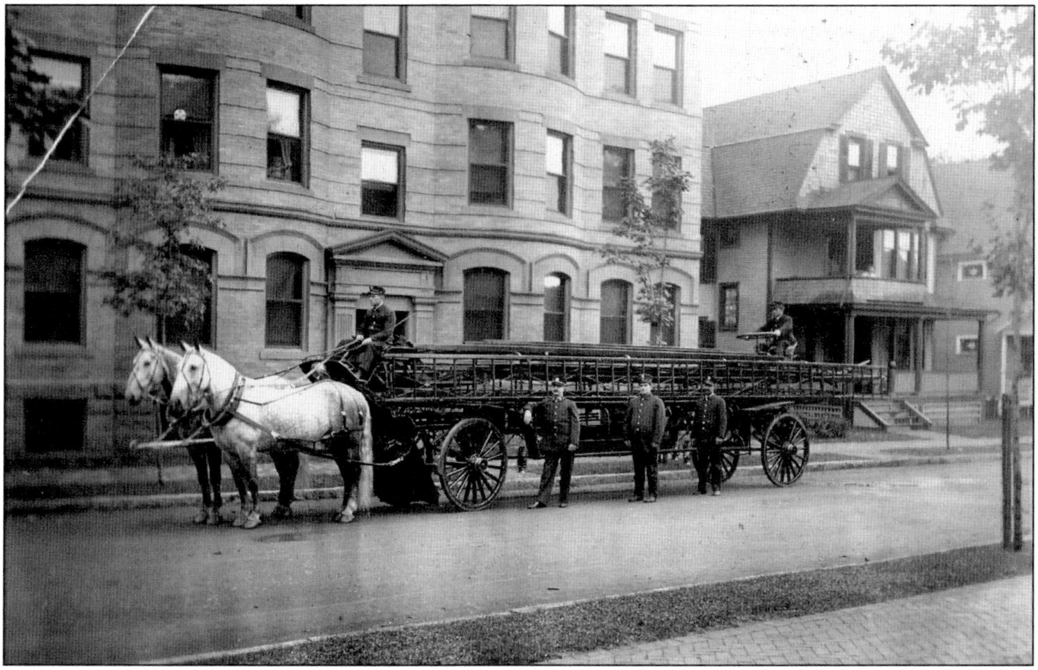

Ladder Company No. 10's city service truck and its firefighters are on Main Street in this 1910 view. (Courtesy Noah Pooler family archives.)

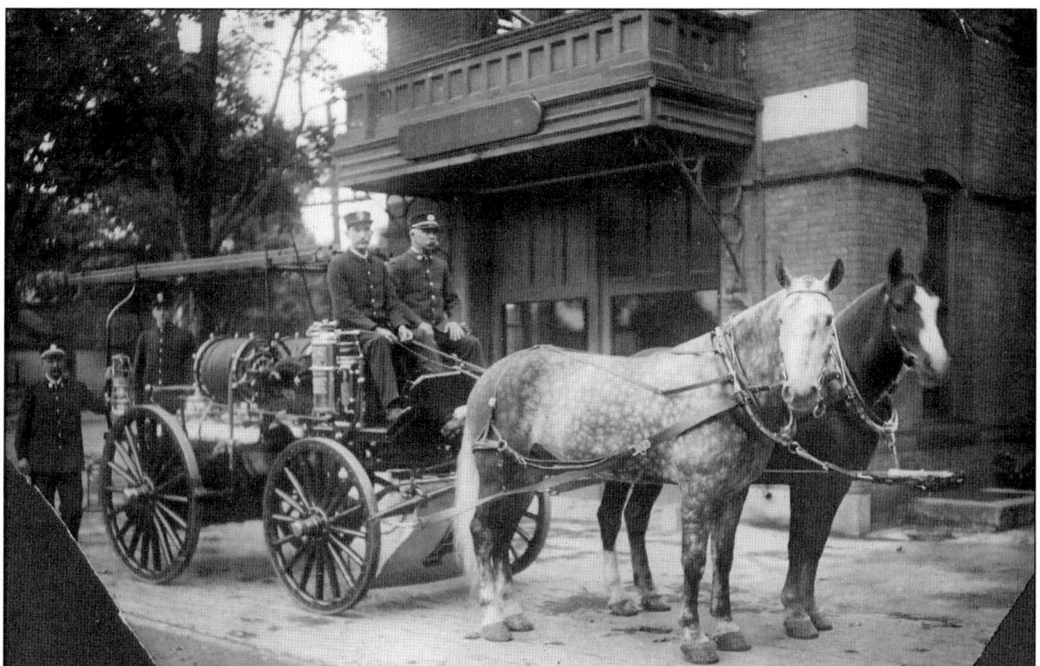

Chemical Company A used this 1894 Holloway chemical wagon and operated out of the Pine Street fire station when it was still a one-bay facility. Chemical Company A later became Squad A and was then reorganized as Engine Company No. 14. (Courtesy Noah Pooler family archives.)

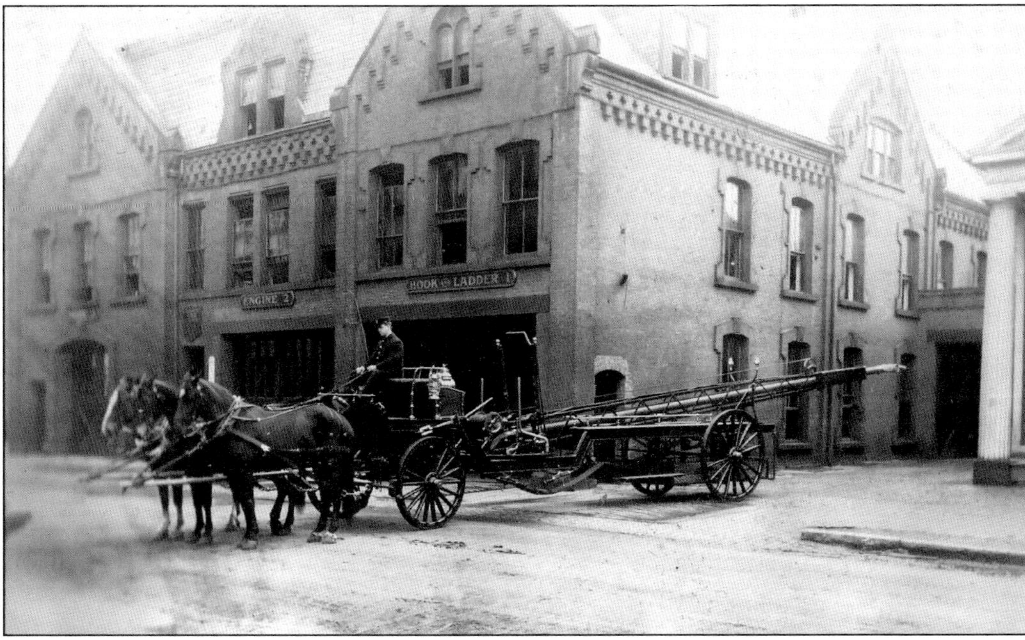

The Pynchon Street fire headquarters, with the department's 1894 Champion water tower built by the Babcock Fire Extinguisher Company, is shown in 1910. After the Glendower Hotel fire in November 1893, the department was convinced that a water tower was badly needed. It was first used at the Gilmore's Opera House fire on April 5, 1897. (Courtesy Noah Pooler family archives.)

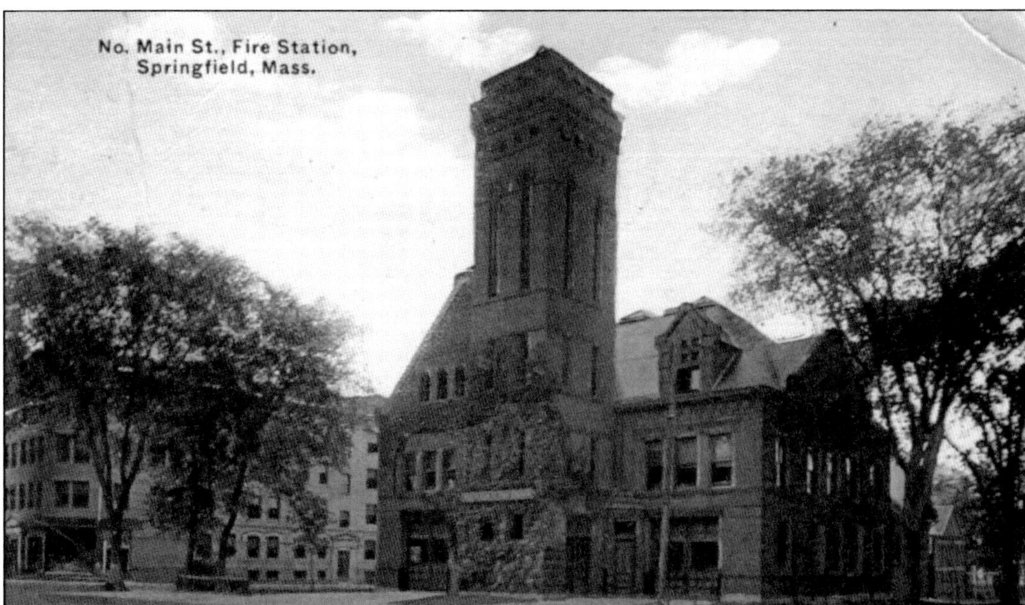

Here is a postcard view of the North Main Street fire station. By 2006, the Pioneer Valley Transit Authority was using the building as an office. (Courtesy Geoffery Neilsen.)

STATIONS AND APPARATUS 1897

ENGINE #1	1871 AMOSKEAG STEAM PUMPER TWO HORSE HOSE WAGON
<u>BOND STREET</u>	
ENGINE #2	1894 AMOSKEAG STEAM PUMPER TWO HORSE HOSE WAGON
CHEMICAL "D"	1894 HOLLOWAY COMBINATION CHEMICAL AND HOSE WAGON
HOOK & LADDER #1	1888 CHAMPION AERIAL LADDER TRUCK
WATER TOWER #1	1894 CHAMPION 65' WATER TOWER
<u>PYNCHONE STREET</u>	
ENGINE #4	1871 AMOSKEAG STEAM PUMPER TWO HORSE HOSE WAGON
<u>WALNUT STREET</u>	
HOSE CO. #5	ONE HAND HOSE CART ONE HORSE HOSE CART
CHEMICAL "F"	TWO HORSE COMBINATION CHEMICAL AND HOSE WAGON
<u>INDIAN ORCHARD</u>	
ENGINE #6	1897 WATEROUS STEAM PUMPER 1887 GLEASON & BAILEY HOSE WAGON
HOOK & LADDER #4	1896 SEAGRAVE CITIES SERVICE LADDER TRUCK
<u>MILL STREET</u>	
HOSE CO. #7	1889 B.F. PIERCE HOSE WAGON
<u>WORTHINGTON STREET</u>	
ENGINE #8	1896 LAFRANCE STEAM PUMPER 1891 FLYNN & KIRBY HOSE WAGON
HOOK & LADDER #2	1871 CITIES SERVICE LADDER TRUCK
<u>WINCHESTER PARK</u>	
HOSE CO. #9	1893 FLYNN & KIRBY HOSE WAGON
HOOK & LADDER #3	1875 CITIES SERVICE LADDER TRUCK
<u>MAIN AND HOOKER STREETS</u>	
CHEMICAL "A"	1894 HOLLOWAY COMBINATION CHEMICAL AND HOSE WAGON
<u>PINE AND CEDAR STREETS</u>	
CHEMICAL "E"	1891 FLYNN & KIRBY CHEMICAL AND HOSE WAGON
<u>OAKLAND AND DICKERSON STREETS</u>	

Here is a list of the stations and apparatus of the Springfield Fire Department in 1897. Over the years, companies were relocated around the city as new companies were organized and new stations opened. For example, Engine Company No. 1 is shown on Bond Street in 1897, but by 1905, it was relocated to Pynchon Street and Engine Company No. 2 was moved to Bond Street. (Courtesy Connecticut Fire Museum archives.)

Some Notable Fires

Springfield Plantation,	June, 1675
United States Armory,	March 2, 1824
Frosts' Block,	October, 1844
Hampden House,	November, 1844
Western Railroad Depot,	March 17, 1847
Ferry Street,	April 12, 1858
United States Armory,	July 2, 1864
Music Hall,	July 24, 1864
Howard Brothers, cotton waste,	February 9, 1866
Wason Car Company,	February 15, 1867
Distillery,	November 20, 1867
Chase Planing Mill,	May 18, 1869
Watch Factory,	April 27, 1870
T. M. Walker,	August 19, 1870
Bemis & Call,	July 27, 1871
Alden's Woolen Mill,	April 12, 1872
Massachusetts Mutual Life Insurance Company,	February 5, 1872
Boston & Albany Round House,	November 10, 1872
Wight Block,	January 5, 1874
Great Fire at Taylor Street,	May 30, 1875
Boston & Albany Car Shop,	September 13, 1876
Indian Orchard Mill,	August 31, 1886
Winthrop Club,	December 2, 1887
Union (newspaper),	March 7, 1888
Glendower Hotel,	November 22, 1893
Gilmore Opera House,	April 5, 1897
City Hall,	January 6, 1905
Highland Baptist Church,	January 3, 1906
Springfield News Company	November 2, 1910
Wason Car Company,	December 19, 1912

Here is a list of some of the notable fires that occurred from 1675 to 1912 in Springfield. (Courtesy Connecticut Fire Museum archives.)

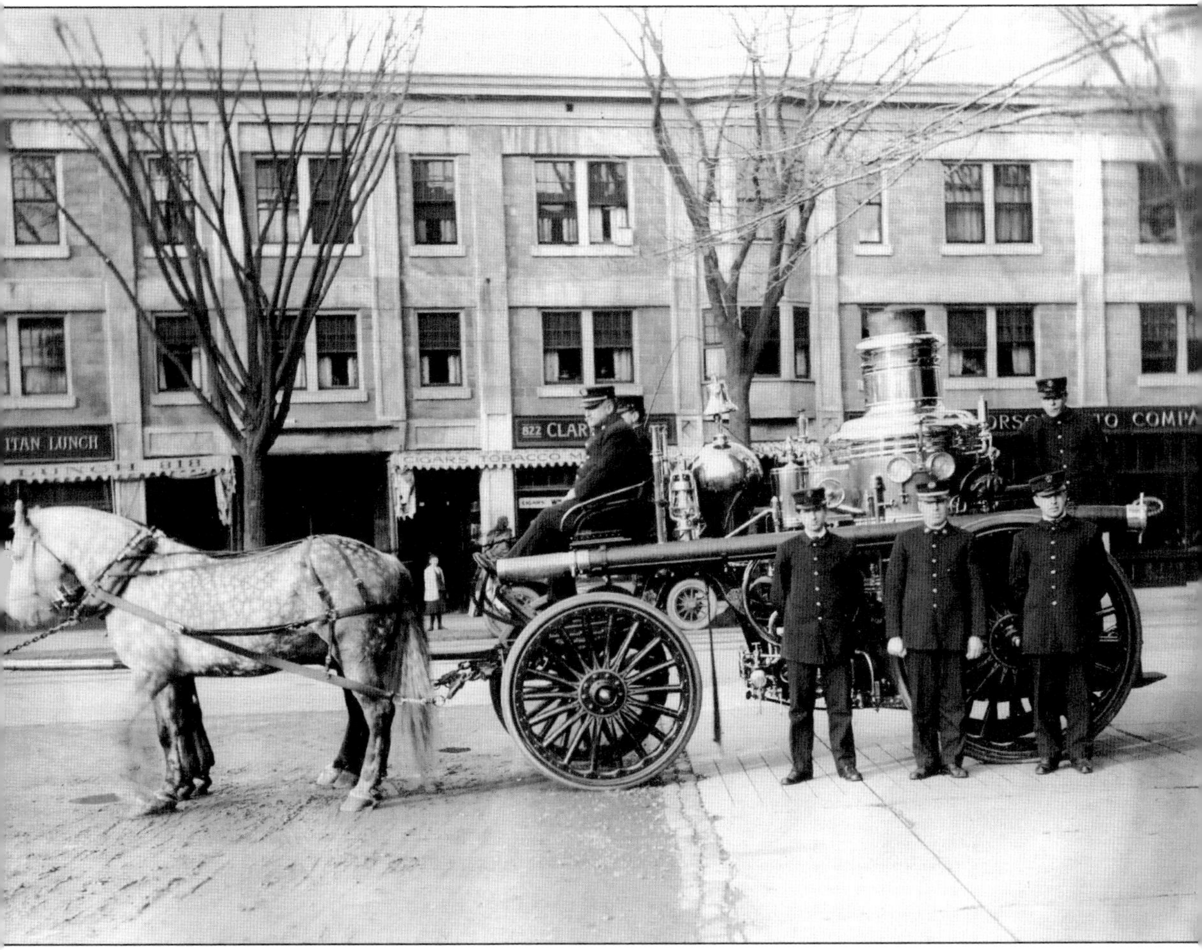

The equipment and members of Engine Company No. 8 are pictured in front of the Winchester Park fire station around 1905. The building shown behind the steamer still stands in 2006. (Courtesy Raymond Pond.)

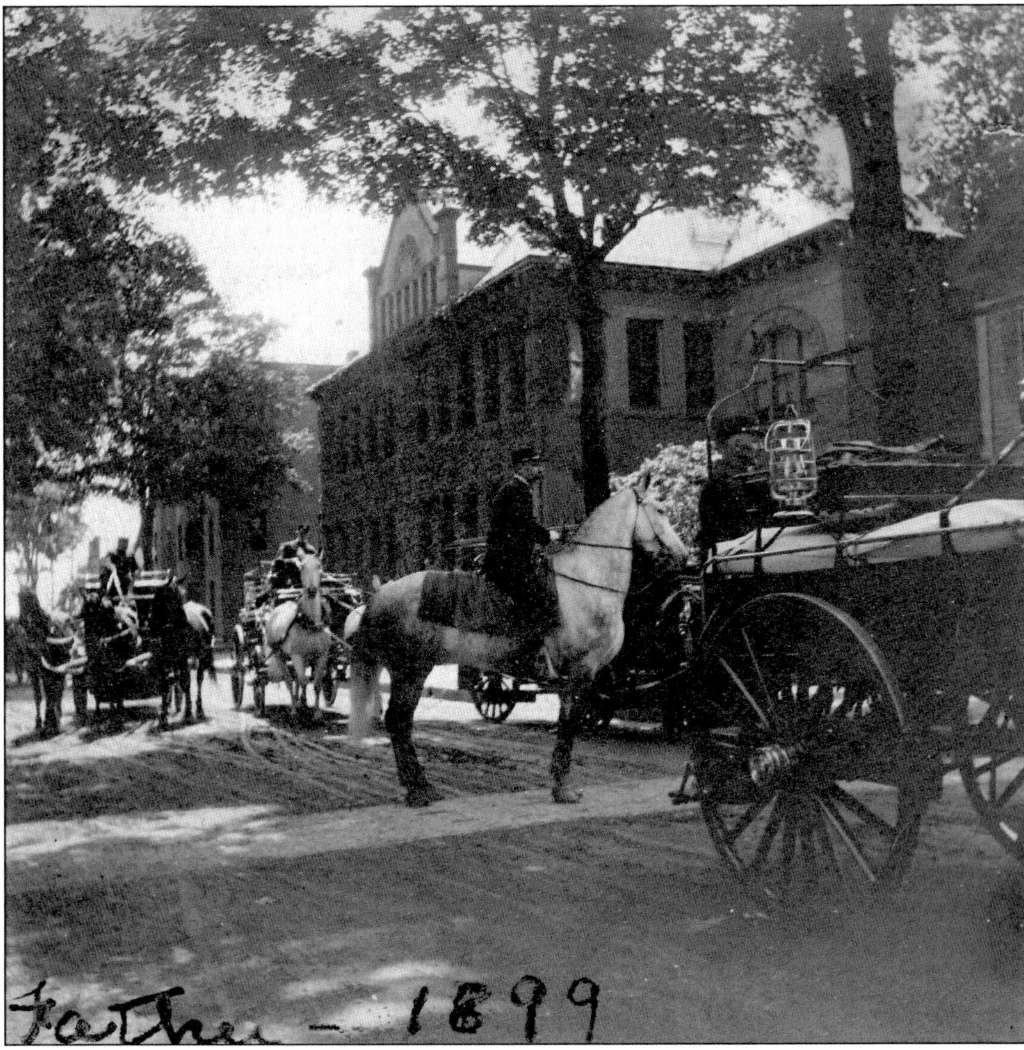

This 1899 view on Pynchon Street shows the chief riding a white horse, most likely for a parade. During the early volunteer days, the chief and all the firemen ran to a fire, but then the chief was given a horse to ride and later was provided with a buggy like the one shown on page 17. (Courtesy Raymond Pond.)

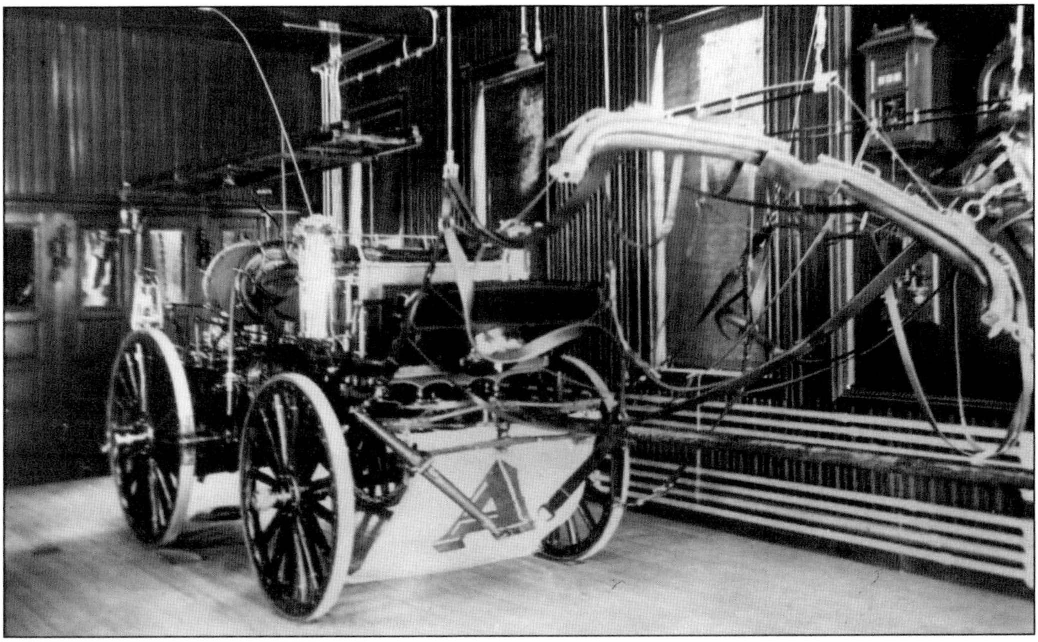

This is an inside view of the wagon of Chemical Company A that shows the quick-connect harness. When an alarm came into the station, a number of events occurred. The stable doors were opened automatically by the fire alarm system, freeing the horses that were trained to walk to their position under the harness. The firefighters climbed on the apparatus, and the driver pulled a cord that released the harness suspended on the ceiling, dropping it onto the horses. Then a fireman snapped the harness buckles to complete the operation. In less than two minutes the wagon would be out the door. (Courtesy Raymond Pond.)

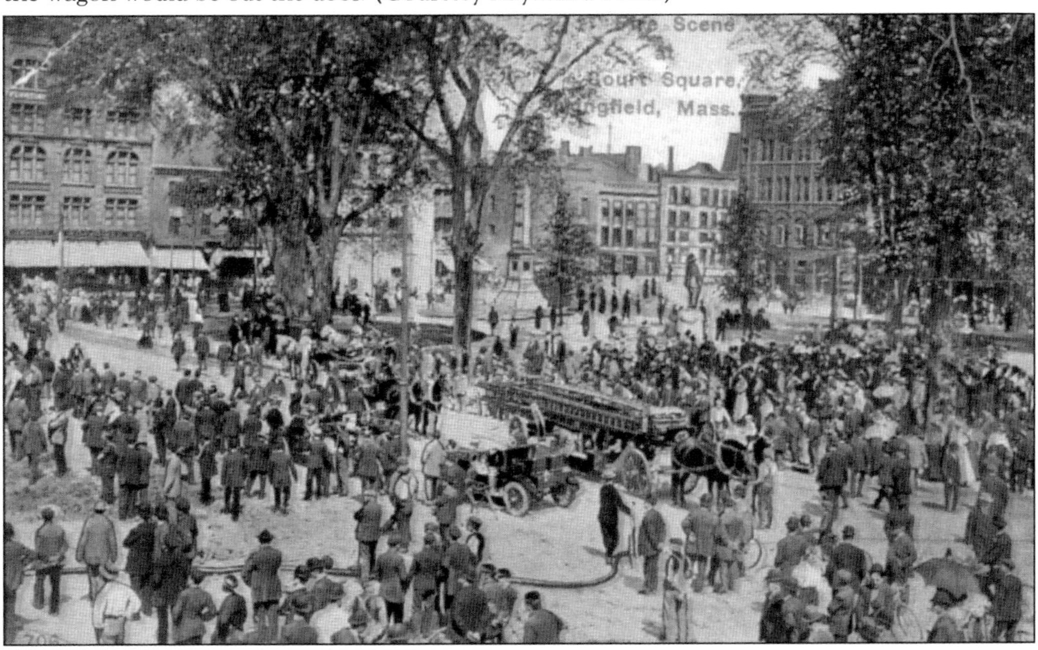

Here is a postcard view of Court Square in 1905 during what appears to be a demonstration by the fire department. (Courtesy Geoffery Neilsen.)

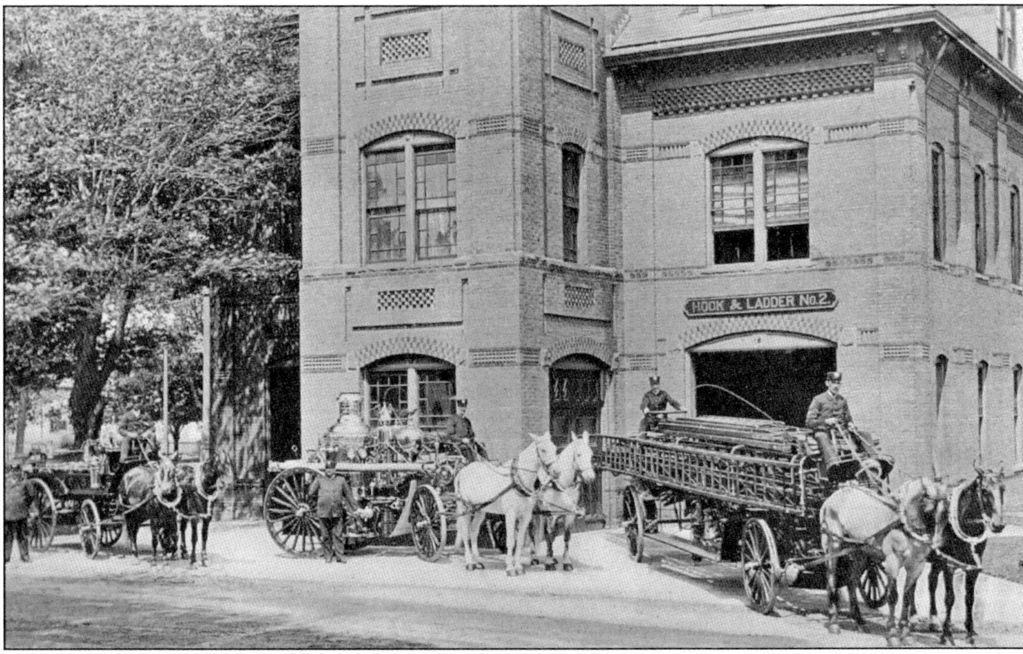

The Winchester Park fire station is shown with the firefighters and their apparatus in 1905. (Courtesy Connecticut Fire Museum archives.)

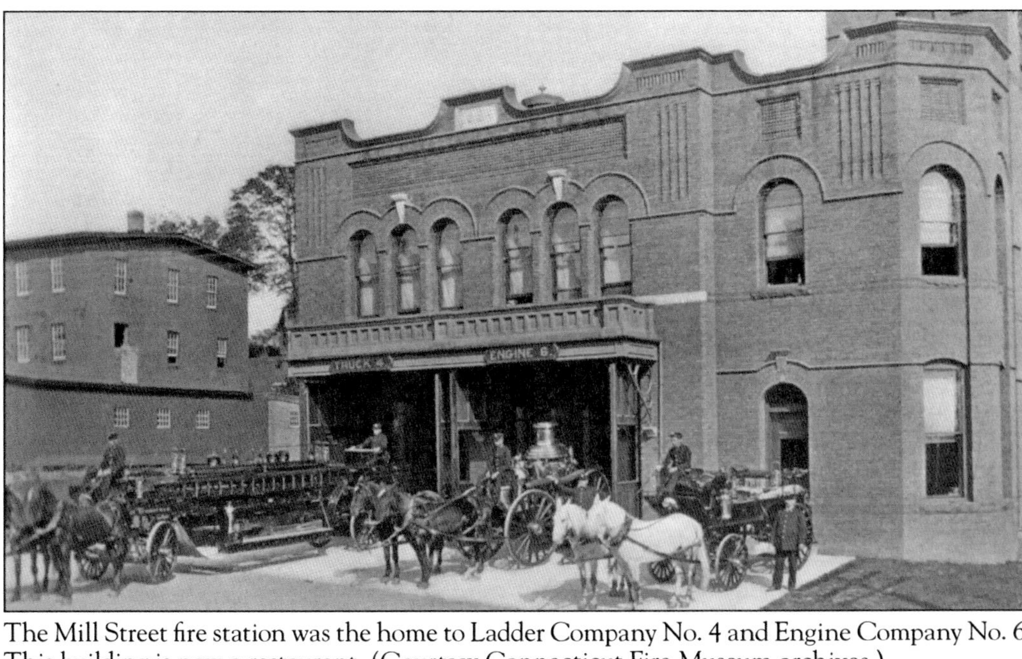

The Mill Street fire station was the home to Ladder Company No. 4 and Engine Company No. 6. This building is now a restaurant. (Courtesy Connecticut Fire Museum archives.)

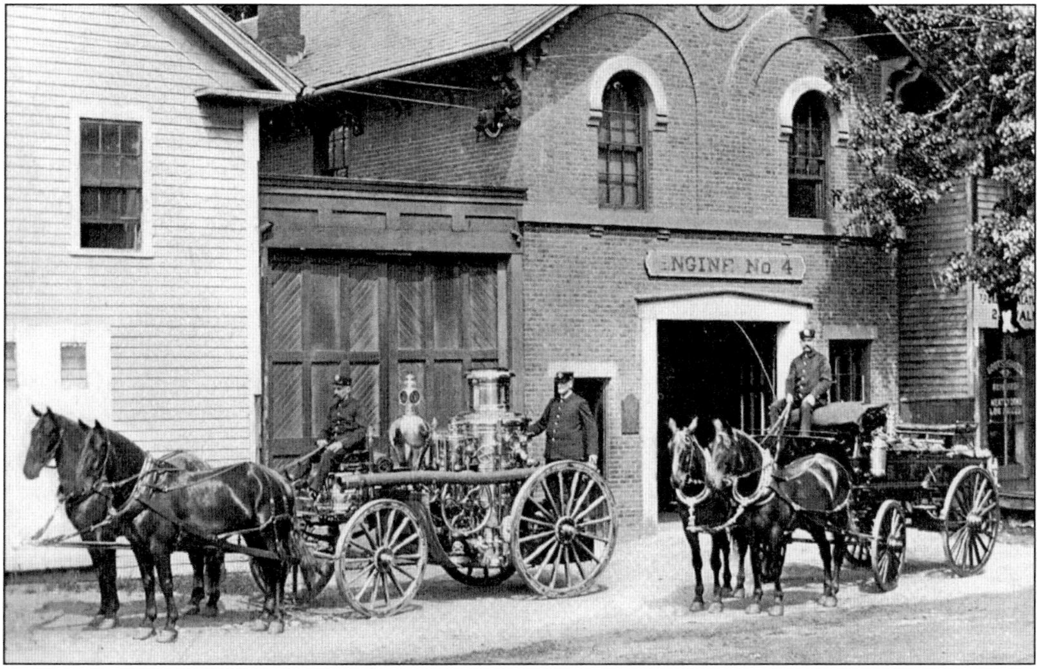

This 1899 view shows the steamer, hose wagon, and crew from Engine Company No. 4. With only one door, the steamer was parked in the rear of the station with the hose wagon near the door. (Courtesy Connecticut Fire Museum archives.)

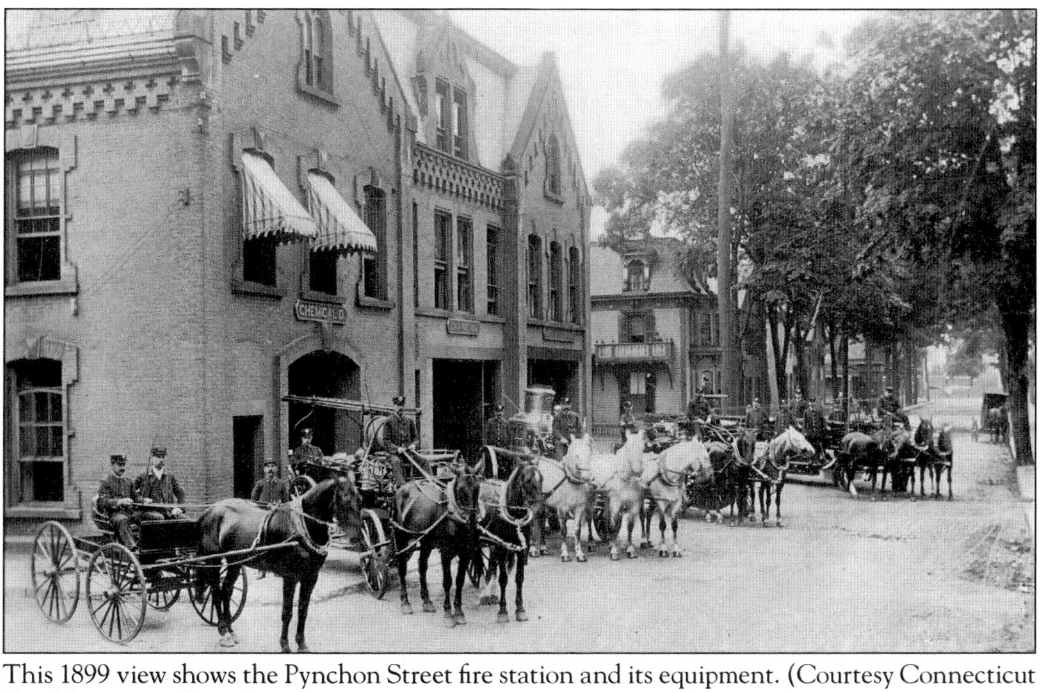

This 1899 view shows the Pynchon Street fire station and its equipment. (Courtesy Connecticut Fire Museum archives.)

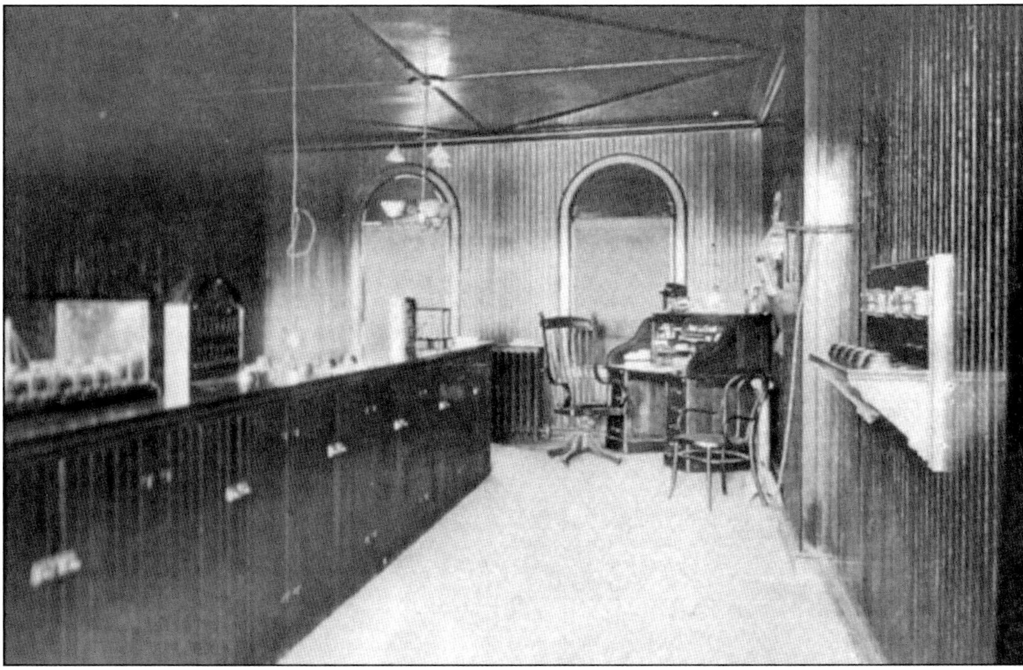

This is an interior view of the fire alarm office on the second floor of the Pynchon Street fire station. All alarms from street fireboxes were received in this room and then resent to the stations that were assigned to that firebox. (Courtesy Connecticut Fire Museum archives.)

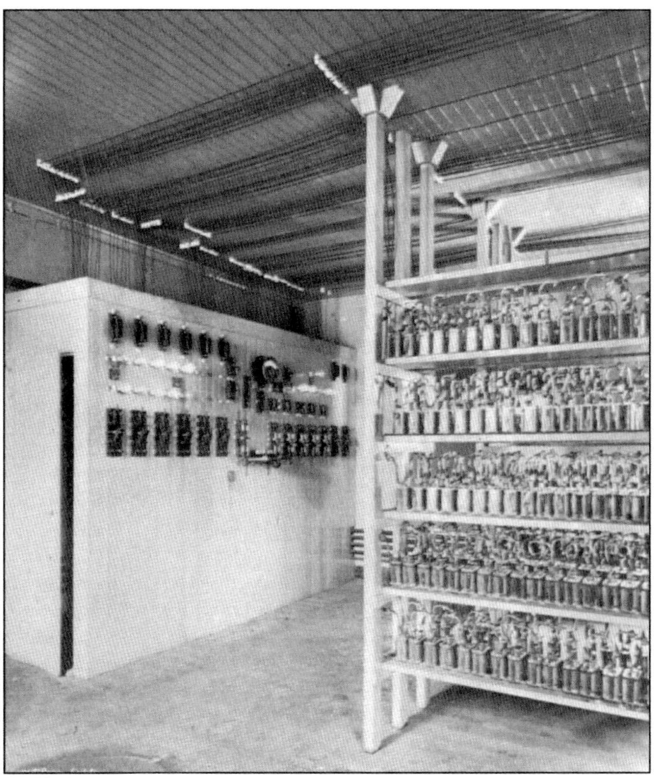

The Gamewell fire alarm system used by Springfield operated on 48 volts of direct current that was supplied to the system by these banks of lead acid batteries. The department had a small steam-driven dynamo to keep the batteries charged. (Courtesy Connecticut Fire Museum archives.)

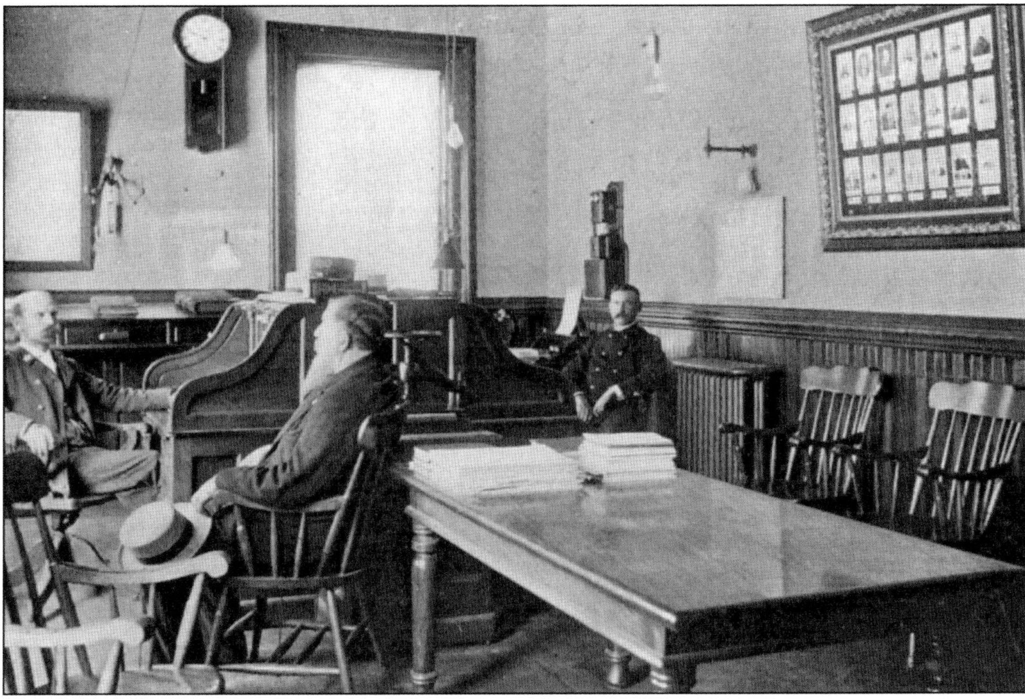

This is a 1905 view of the chief's office on the second floor of the Pynchon Street fire station. (Courtesy Connecticut Fire Museum archives.)

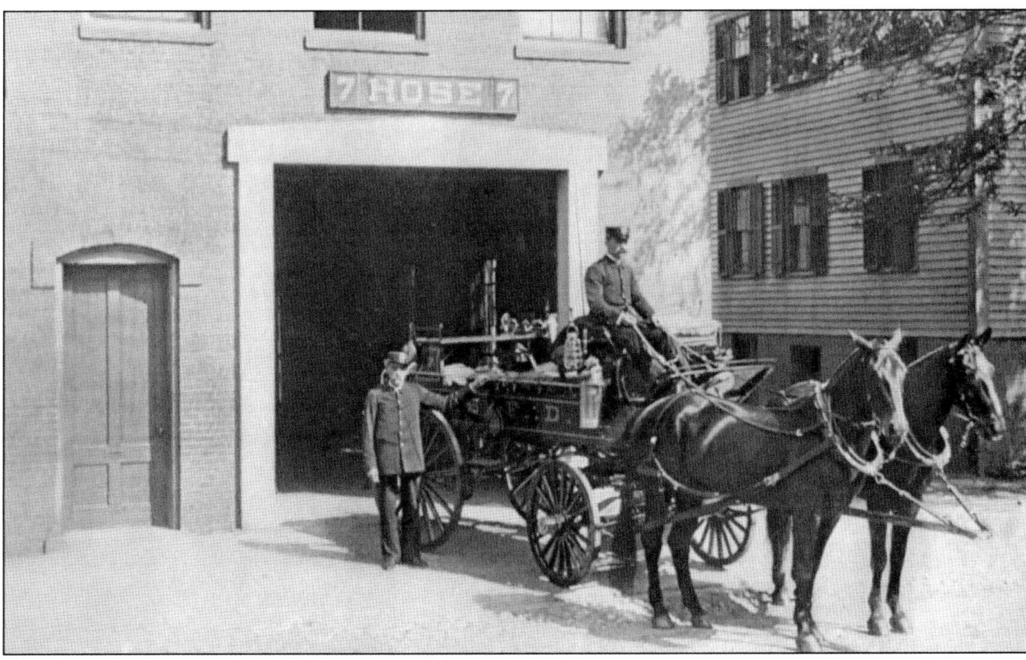

The Worthington Street fire station was the home of Hose Company No. 7 in 1905. Springfield had a very good water system, so some of the companies did not operate with a steamer as the pressure from the hydrants was enough to extinguish a small fire. (Courtesy Connecticut Fire Museum archives.)

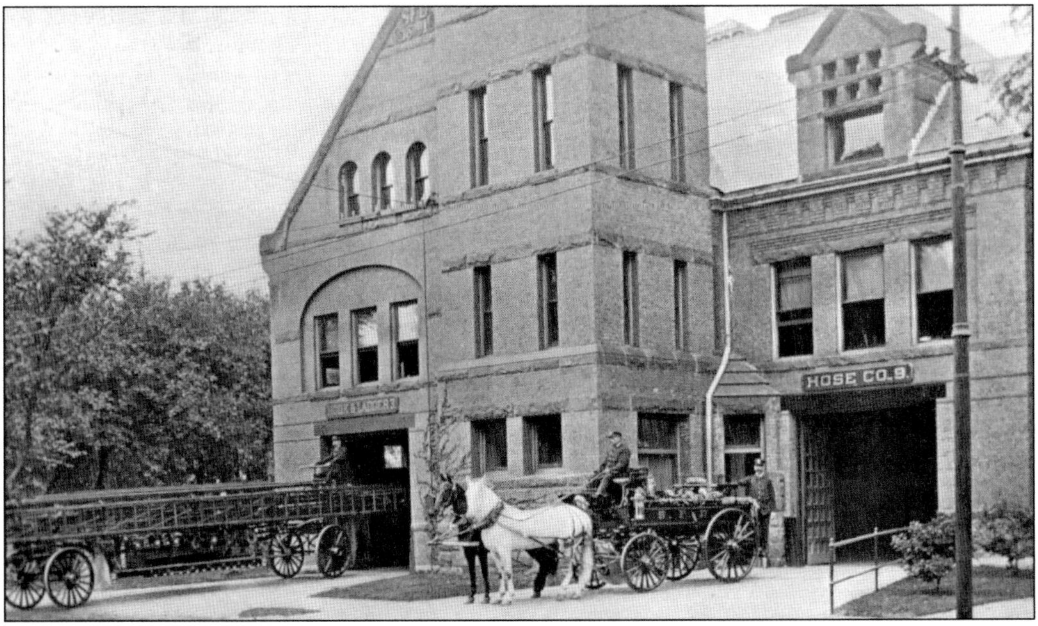

The fire station at 2808 North Main Street was home to Ladder Company No. 3 before the company was relocated to the Oakland Street fire station. The North Main Street building still stands today and is owned by Pioneer Valley Transit Authority and used as an office building. A new fire station was built a few blocks south at 2729 North Main Street. (Courtesy Connecticut Fire Museum archives.)

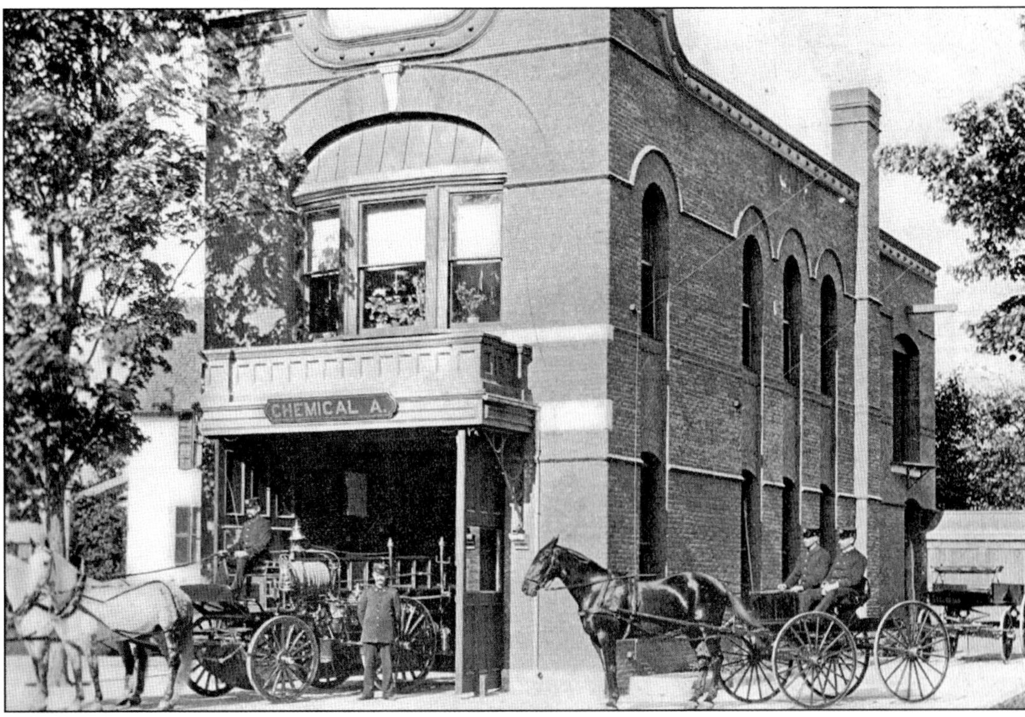

The fire station at 147 Pine Street was originally built as a one-bay station to house Chemical Company A. (Courtesy Connecticut Fire Museum archives.)

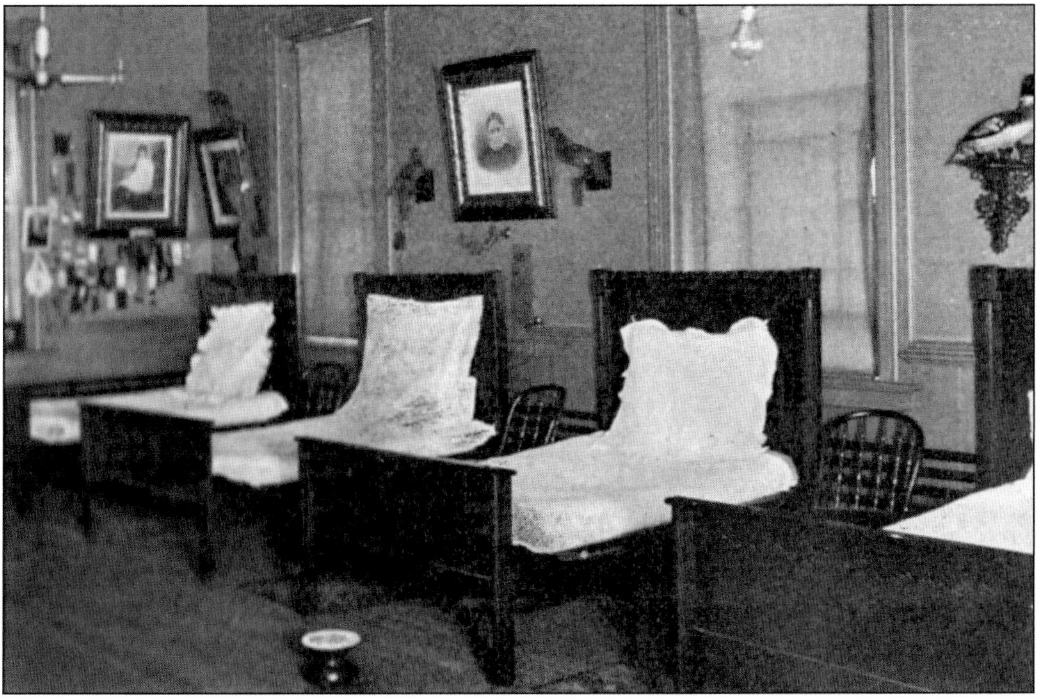

Pictured here are typical sleeping quarters at a Springfield fire station in the early 1900s. (Courtesy Connecticut Fire Museum archives.)

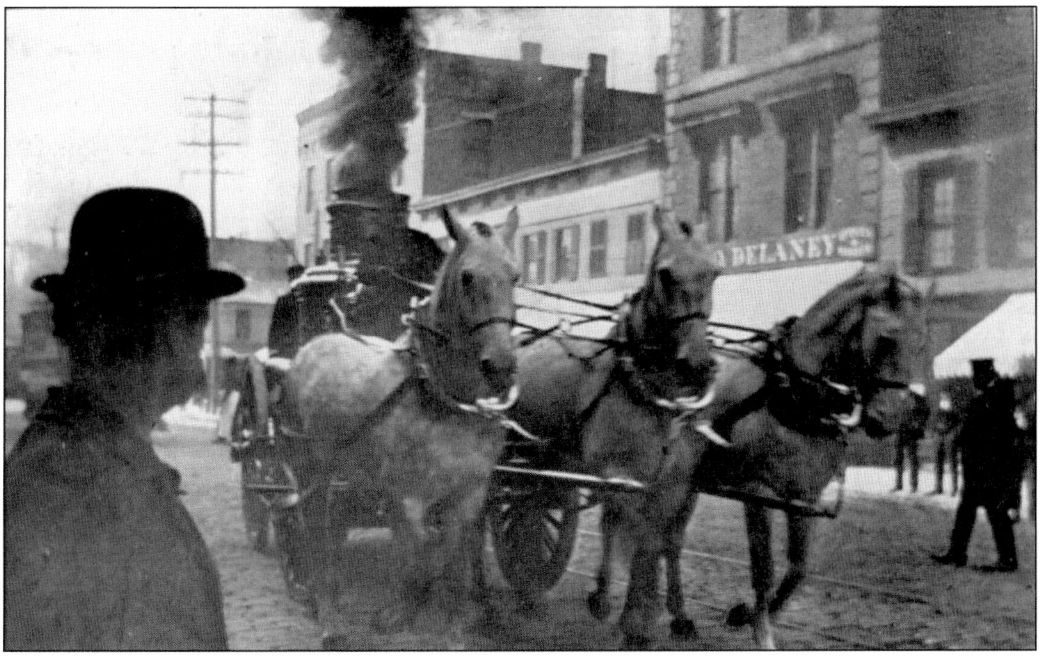

With horses and men hard at work, this three-horse hitch is hauling the 1894 Amoskeag steam pumper from Engine Company No. 2 while responding to an alarm in 1897. The engineer already has steam up, and the engine will be ready to pump water as soon as it arrives at the fire scene. (Courtesy Connecticut Fire Museum archives.)

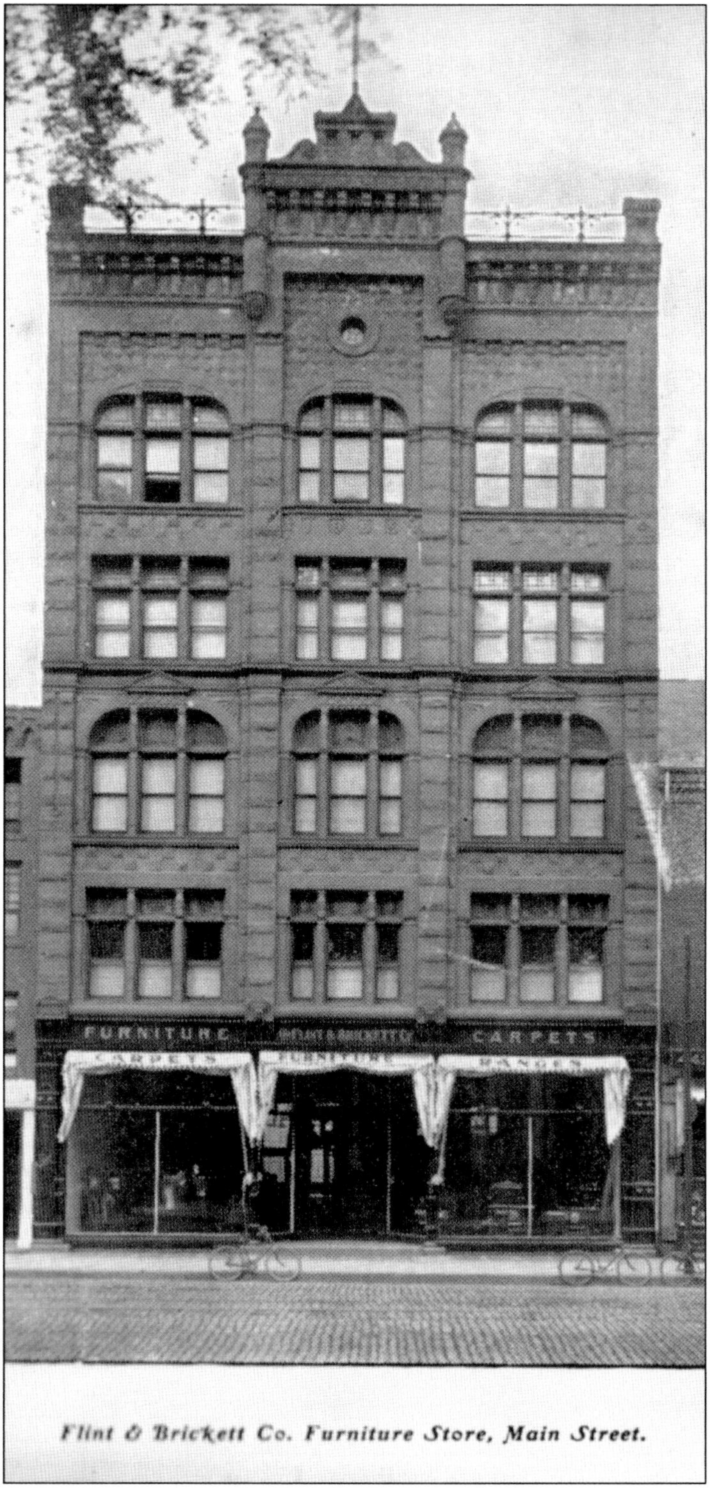

This view shows the Flint and Becket building on Main Street before it was destroyed by fire in 1914. (Courtesy Connecticut Fire Museum archives.)

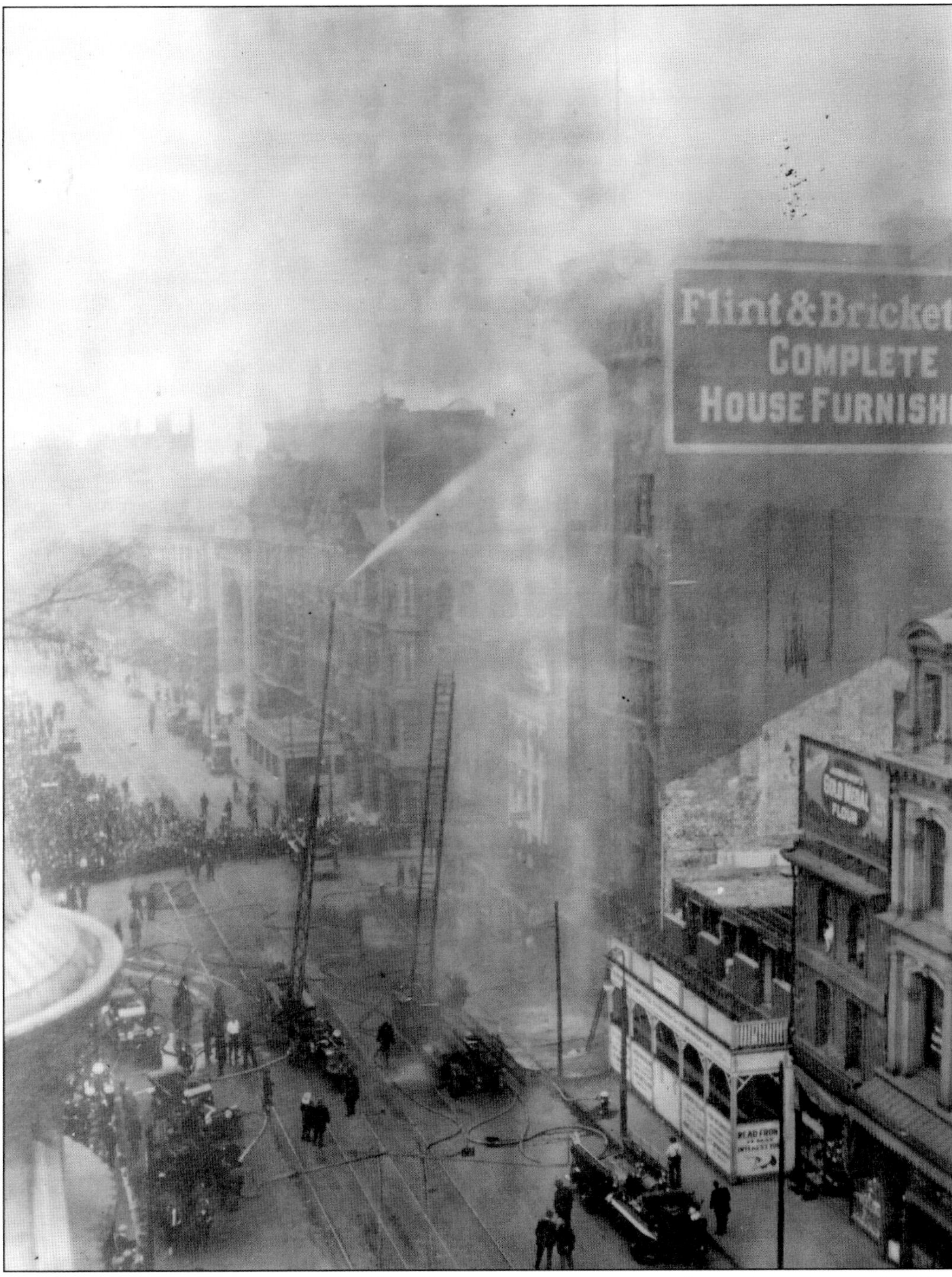

The Springfield Fire Department's 1894 Champion water tower is shown in action at the Flint and Becket fire on October 6, 1914. (Courtesy Connecticut Fire Museum archives.)

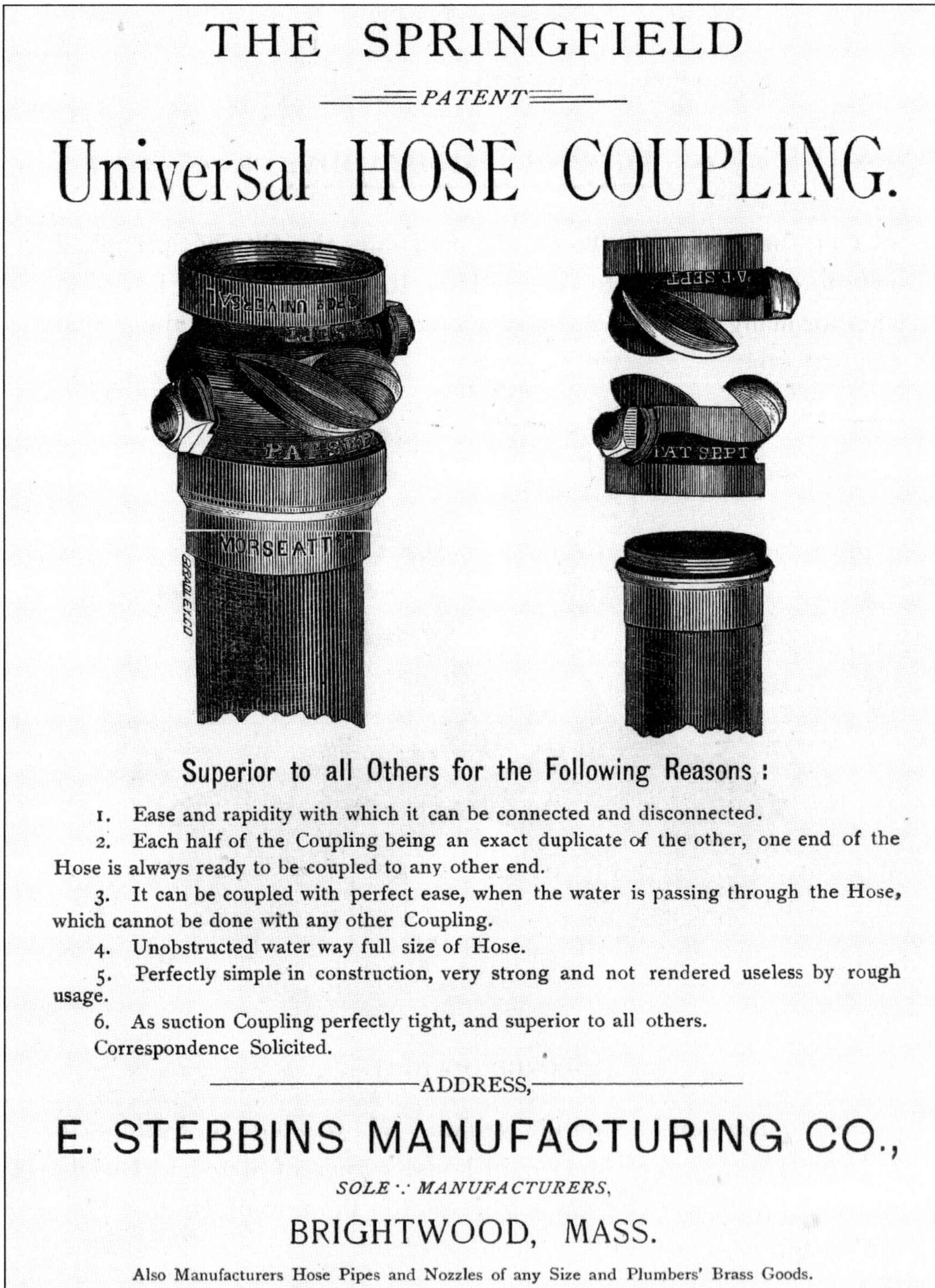

The E. Stebbins Manufacturing Company, located in the Brightwood section of Springfield, designed and marketed a quick disconnect fire hose coupling. This was very similar to the German-designed Storz couplings used today by many fire departments in Europe and North America. (Courtesy Connecticut Fire Museum archives.)

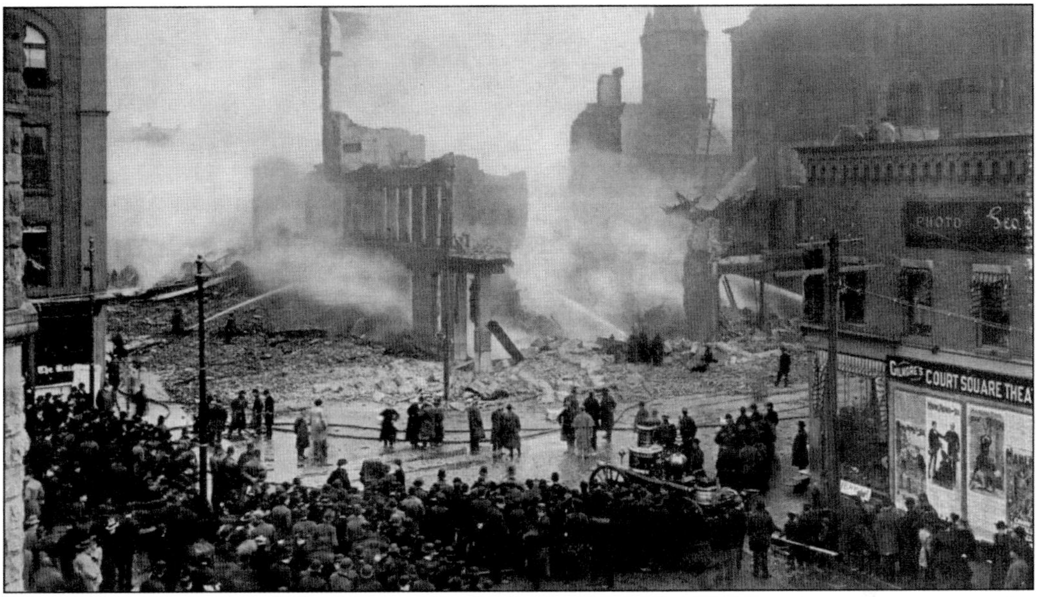

Just after midnight on November 22, 1893, a fire alarm was received from box 41 for a fire at the J. K. Dexter block on Worthington Street. Soon after, the fire spread to the Glendower Hotel at the corner of Worthington and Main Streets. The Union Building across the street also caught on fire, but the firefighters were able to contain that blaze. The Worcester, Hartford, Holyoke, and Chicopee fire departments provided assistance to Springfield. The loss was over $450,000, an enormous sum of money for the time. (Courtesy Connecticut Fire Museum archives.)

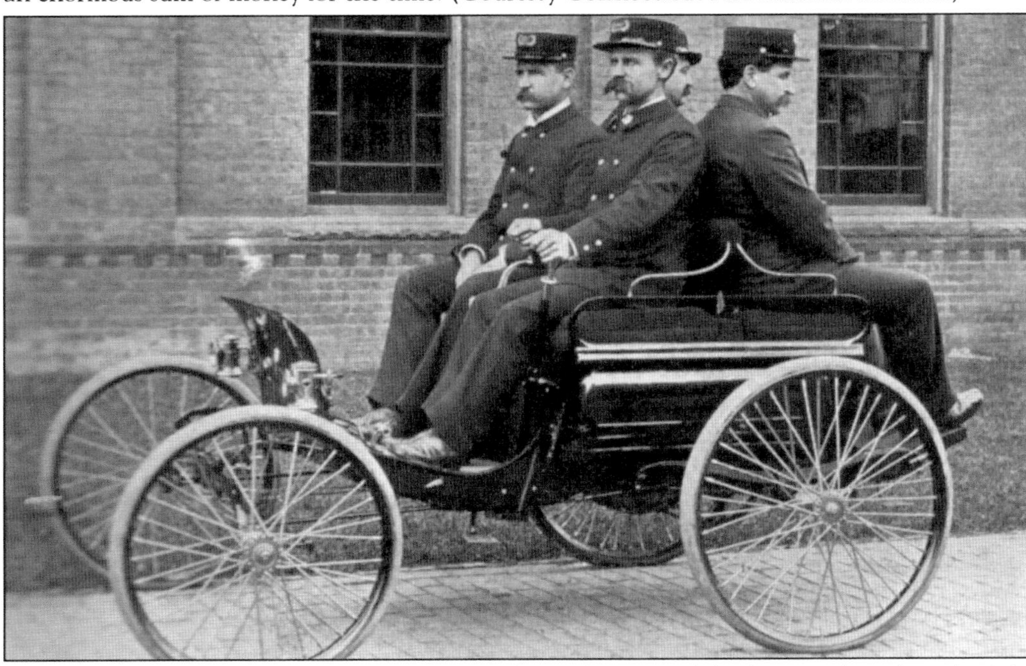

In 1894, the Springfield Fire Department used this early automobile that may have been built by Charles and J. Frank Duryea. The Duryeas built their first automobile in 1893 in Springfield, and the Duryea Automobile Company was the first American firm to build a gasoline automobile. (Connecticut Fire Museum archives.)

✠ ✠ ✠ ✠ Fire Alarm Telegraph. ✠ ✠ ✠ ✠

LOCATION OF SIGNAL BOXES AND KEYS.

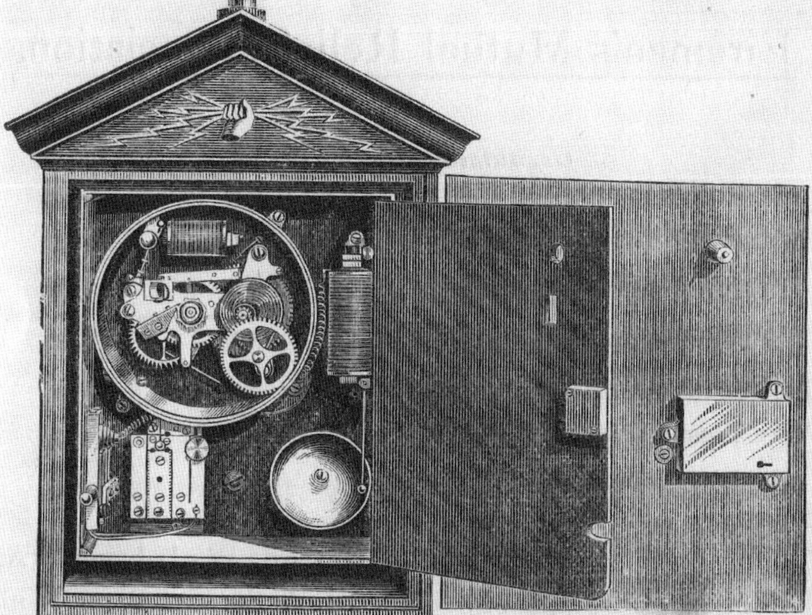

GAMEWELL FIRE ALARM BOX.

3. B. & A. R. R. depot; key with the watchman, at the Massasoit House, and on box.
4. Corner Main and Franklin; key at Stable, 107 Main street, and on box.
5. Corner Main and Carew; key at Horse R. R. Office, at Engine House, Bond street, and on box.
6. Corner Chestnut and Carew; key at Rev. H. Cooley's, and Horace Wheeler's, and on box.
7. Corner Chestnut and Liberty; key at Day & Jobson's Office, at J. Moore's, and on box.
8. Corner Dwight and Taylor; key at Robinson's lumber yard, office of planing mill, and on box.

This is a list of some of the firebox locations in Springfield in 1888. During this era, the boxes were locked and it was necessary to find the "responsible person" who held the key to the box. (Courtesy Connecticut Fire Museum archives.)

FIRE SERVICE OF SPRINGFIELD. 81

12. Corner of Union and School; key at C. P. Dean's, and on box.

13. Corner State and Byers; key at U. S. Guard House.

14. Corner Worthington and Spring; key at hose house, Worthington street, Hatfield's house on the corner, T. Hamlin's, and on box.

15. Corner State and Walnut; key at Allen's meat market, Sexton's stable, engine house, and on box.

16. Corner Orleans and Tyler; key at office of watch factory, and watchman.

17. Corner State and Hancock; key at H. W. Ellis's, E. D. Stock's, T. L. Haynes', and on box.

18. Corner Bay and Buckingham; key at E. H. Lathrop's, 166 Buckingham street, and on box.

19. Almshouse, Boston road; key at Almshouse.

21. Corner Main and State; key at Webber's drug store, at Homer Foot & Co.'s, and on box.

23. Corner Main and Central; key at Waterman's store, M. Hancock's, and on box.

24. Maple street, between Mulberry and Central; key at G. R. Townsley's, H. M. Morehouse's, Avon Place, and W. F. Callender's.

25. Corner Central and Pine; key at Bennett's store, and on box.

26. Corner Maple and Mill, Ames' store; key at K. Dorne's, and G. R. Dickinson's.

27. Corner Main and Mill; key at John Hart's, at Hose 6 house, watchman at Alden's mill, and on box.

28. Corner Main and Pynchon streets; key at Draper's drug store, Haynes Hotel, and on box.

29. Corner Harrison avenue and Dwight street; key at Morgan Envelope Co.'s, at stable on Dwight street, and on box.

Here is another list of firebox locations in 1888. A fire alarm system was first installed in Springfield in 1868 before the advent of the telephone, which was not to be perfected for another 10 years. The familiar red street box was the only way to turn in an alarm unless one wanted to run to the nearest fire station. (Courtesy Connecticut Fire Museum archives.)

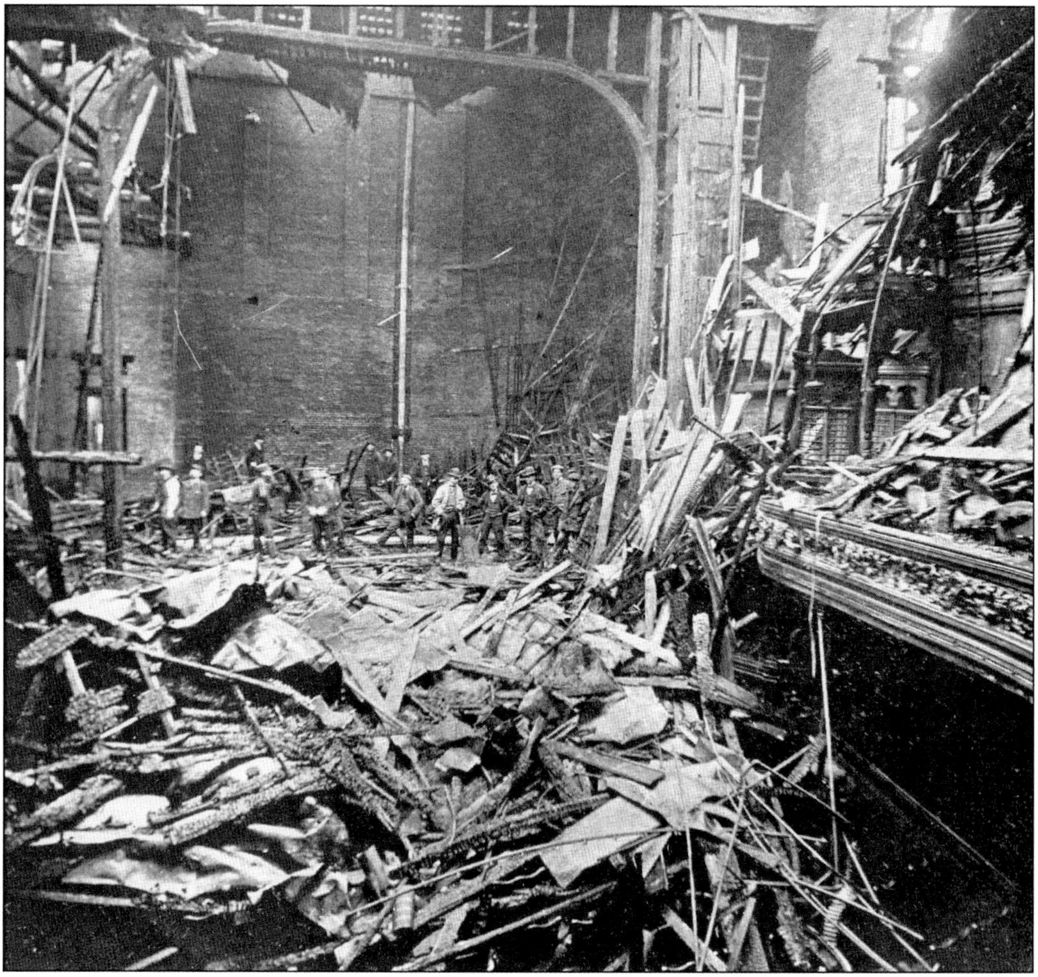

On April 5, 1897, firebox 20 was struck at 11:30 p.m. for a fire that broke out in the old Gilmore Opera House and resulted in the nearly complete destruction of the building. The fire eventually went to two alarms, requiring almost all the equipment of the Springfield Fire Department. It was ironic because all the fire alarm and police circuits for the city actually passed through the building in a wooden conduit and, as a result of the fire, the city lost the use of most of the fire alarm boxes. This was the first fire for which the department used the new Champion water tower. (Courtesy Connecticut Fire Museum archives.)

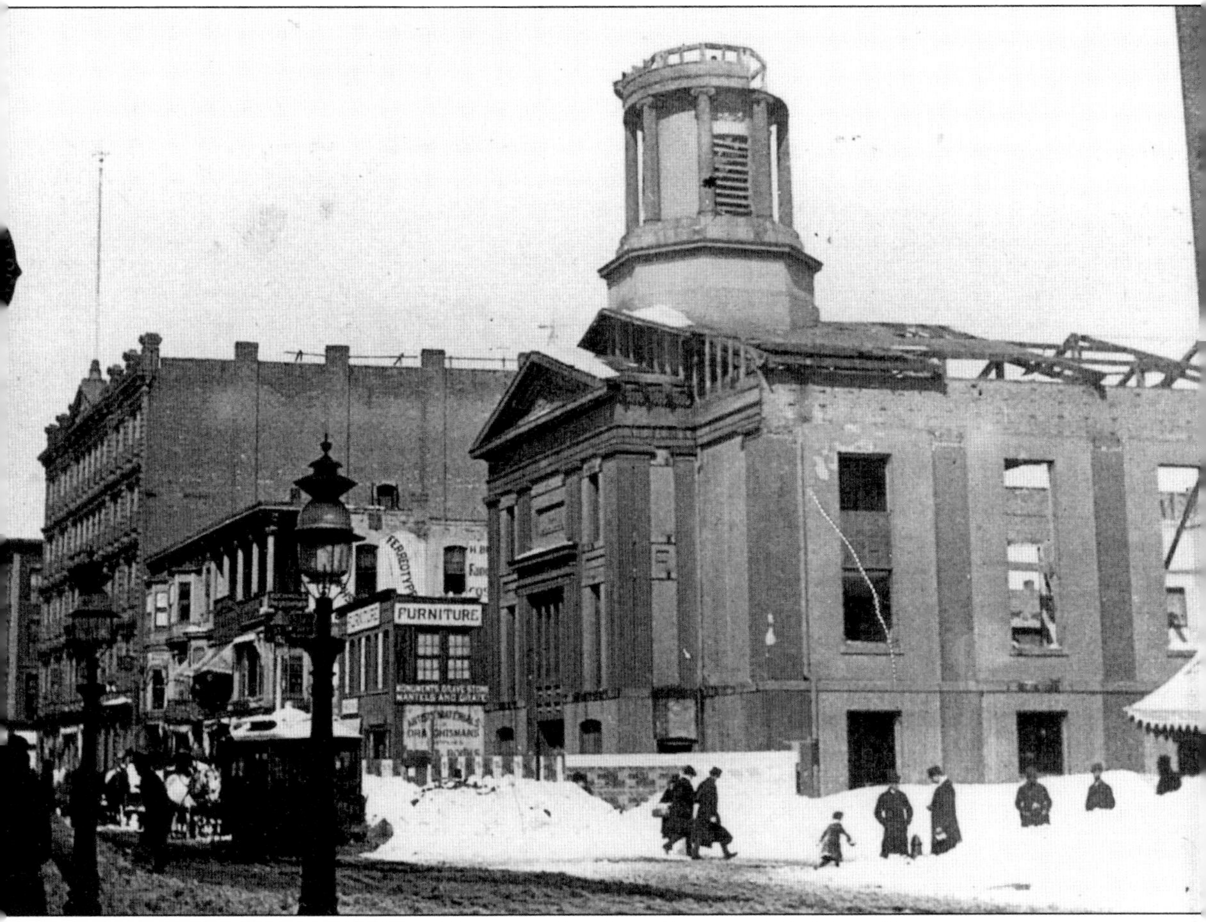

On January 3, 1906, the Highland Baptist Church at State and Stebbins Streets was entirely destroyed by fire. This view shows the fire scene on January 4, 1906. (Courtesy Connecticut Fire Museum archives.)

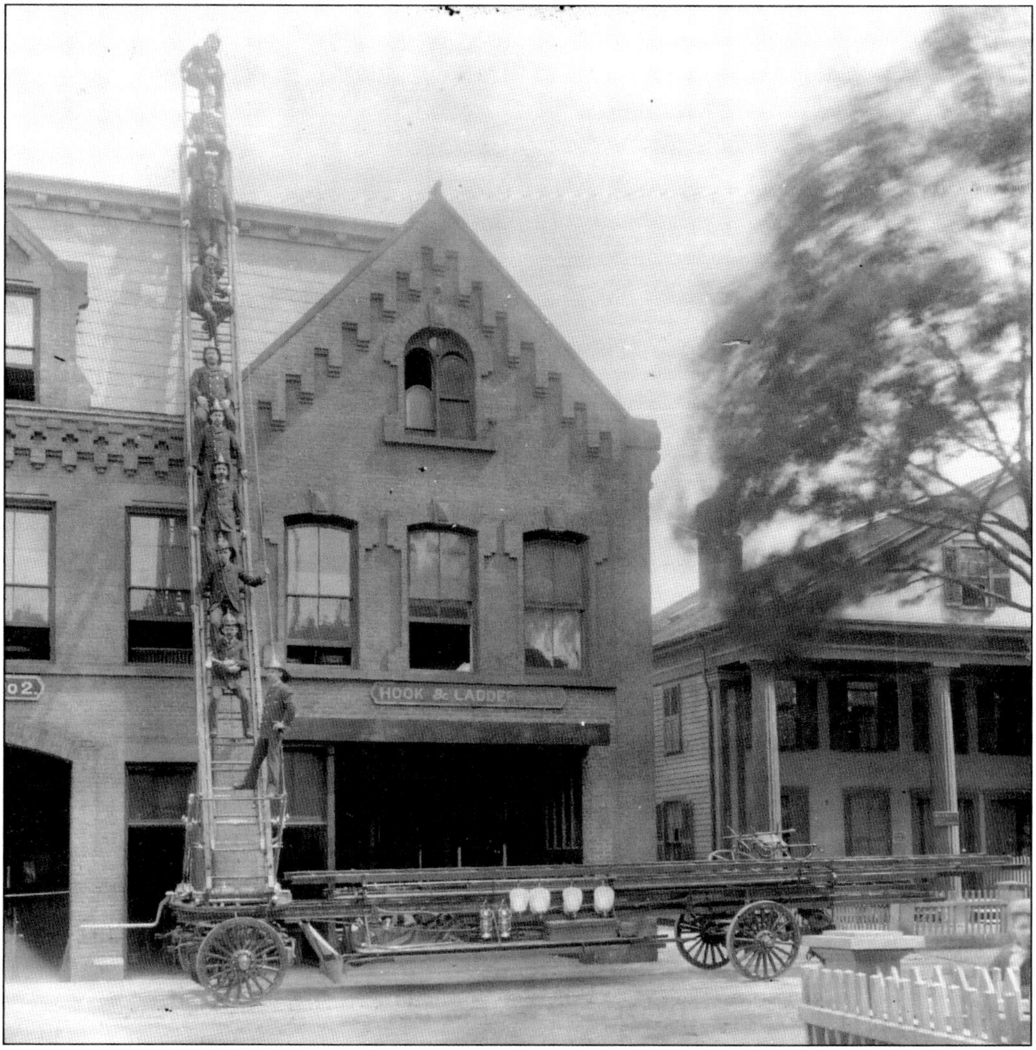

In this photograph, Ladder Company No. 1 is in front of the Pynchon Street fire station with its aerial ladder raised and the company posing for its picture. (Courtesy Noah Pooler family archives.)

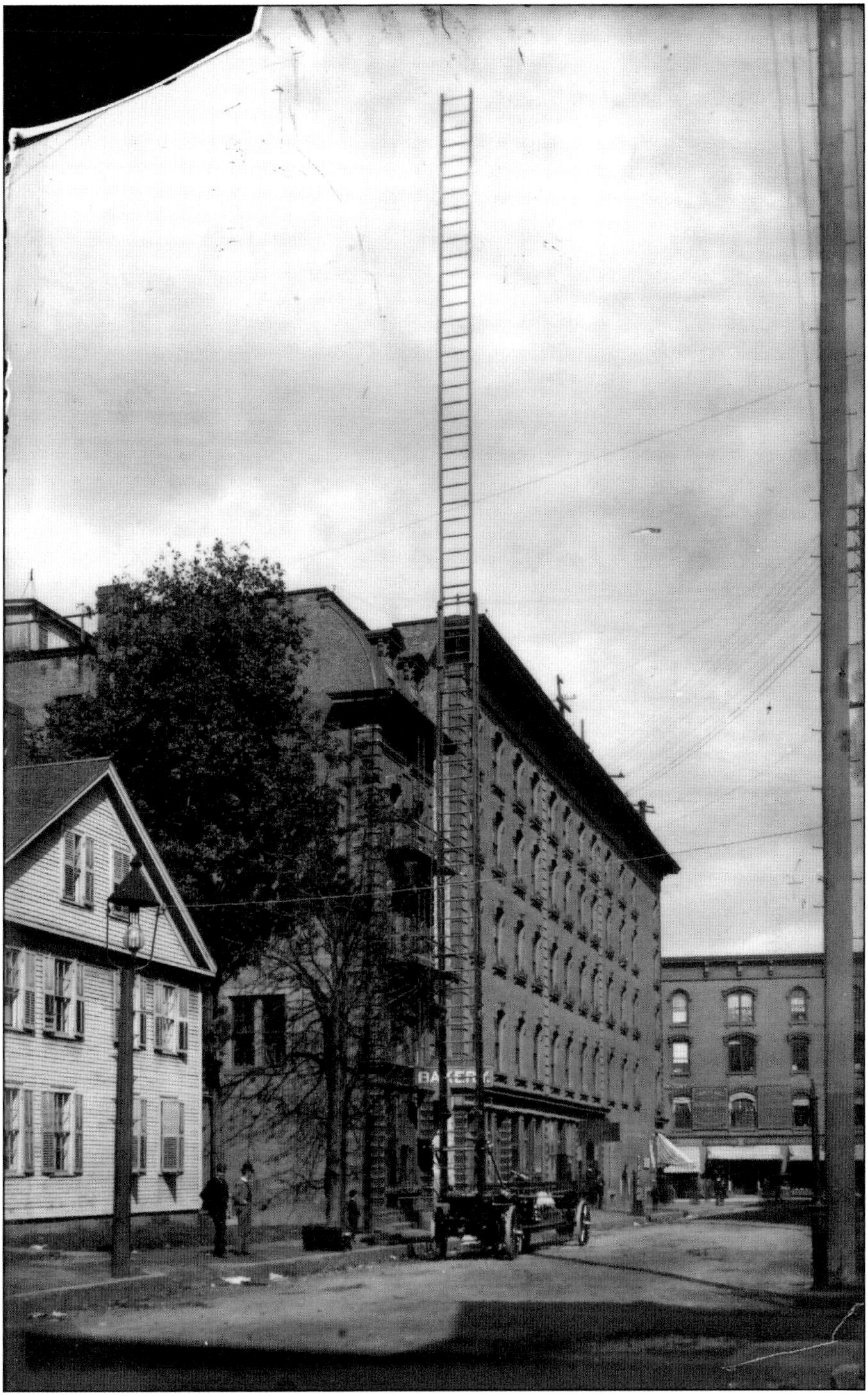

Here is another view of Ladder Company No. 1 with the aerial ladder at its full 85-foot height. In addition to the aerial ladder, the truck carried over 200 feet of ground ladders that ranged in height from 16 feet to 50 feet. These ladders were all raised by hand, and with the 50-foot ladder weighing over 500 pounds, a large number of firefighters were needed on a ladder truck. (Courtesy Noah Pooler family archives.)

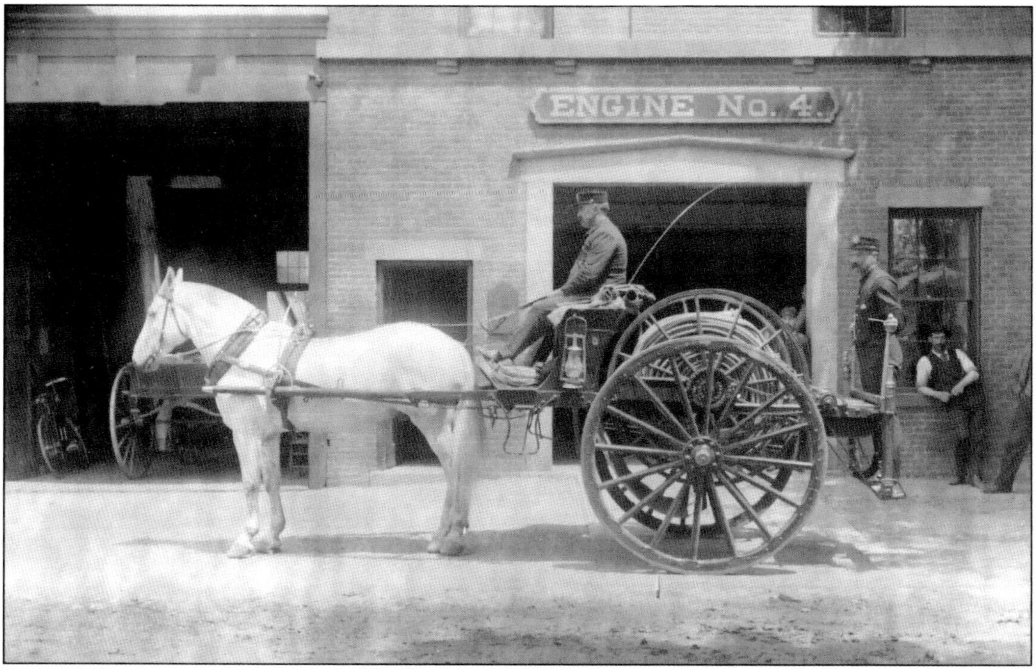

Engine Company No. 4 ran this little two-wheel hose cart until it was replaced by a larger four-wheel hose wagon around 1898. (Courtesy Noah Pooler family archives.)

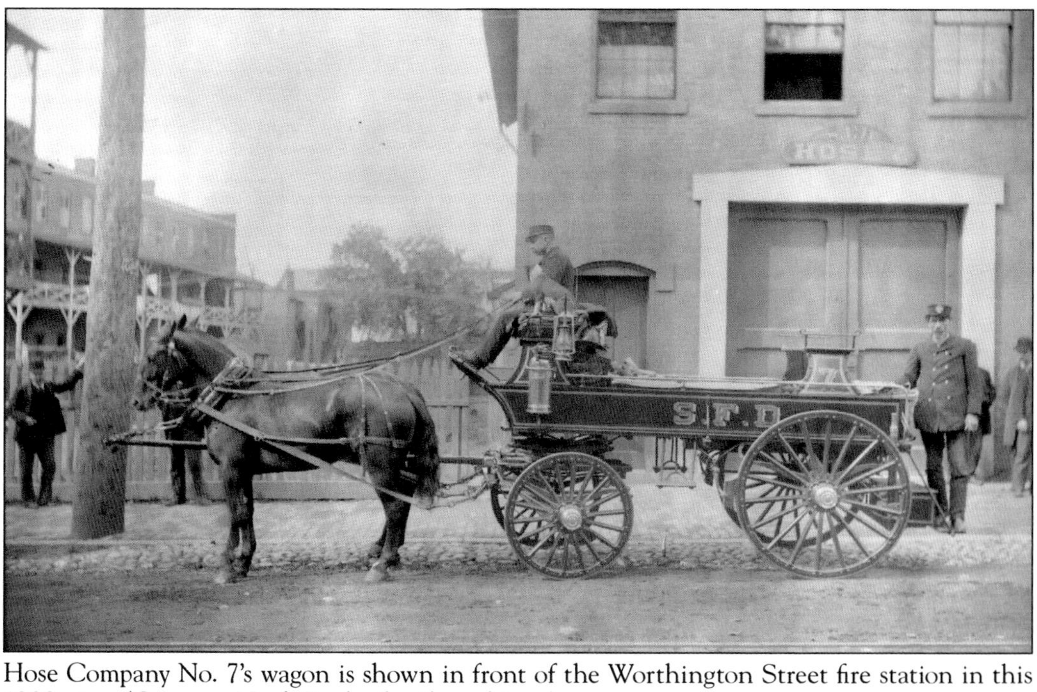

Hose Company No. 7's wagon is shown in front of the Worthington Street fire station in this 1900 view. (Courtesy Noah Pooler family archives.)

Three

Springfield Motorized

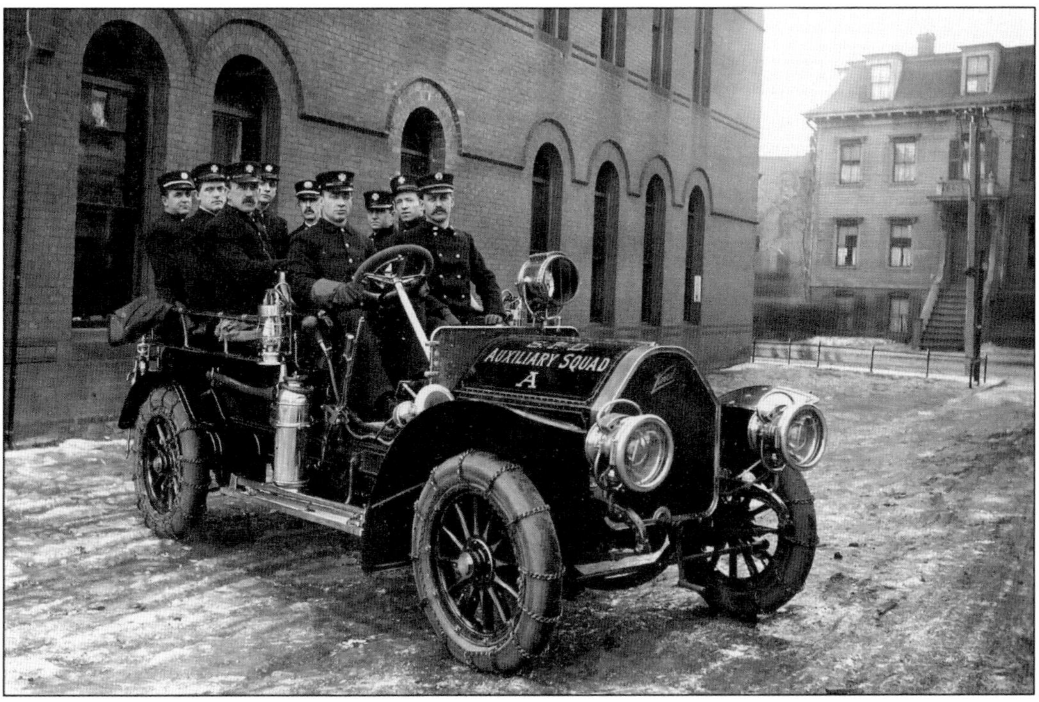

The firefighters of Squad A are shown at the Chestnut Street firehouse with their 1906 Knox apparatus. In 1906, the department had 80 horses worth $200 each available to pull the steamers and hose wagons. By 1916, the department was able to stop using all of the remaining horses with the complete motorization of the department. (Courtesy Noah Pooler family archives.)

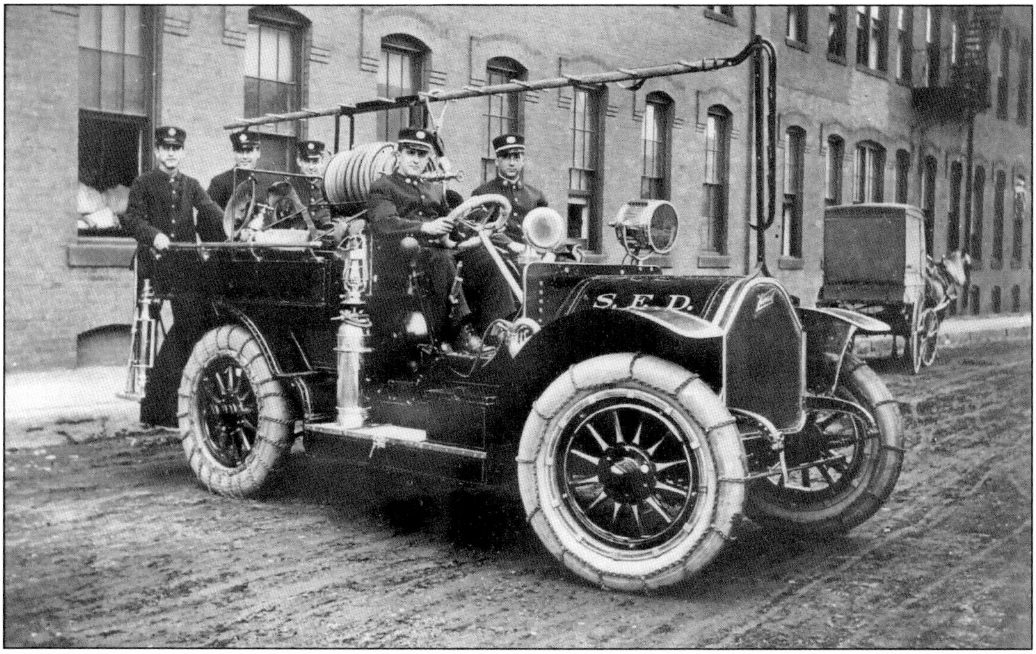

The members of Hose Company No. 7, with their Knox chemical and hose wagon, are shown in this 1908 view. It is interesting to note that the truck does not have any headlights, just the single spotlight on the dash. The tires chains are not for snow and ice but for traction on the dirt and gravel streets. Early-style pneumatic tires did not have any thread on them. (Courtesy Connecticut Fire Museum archives.)

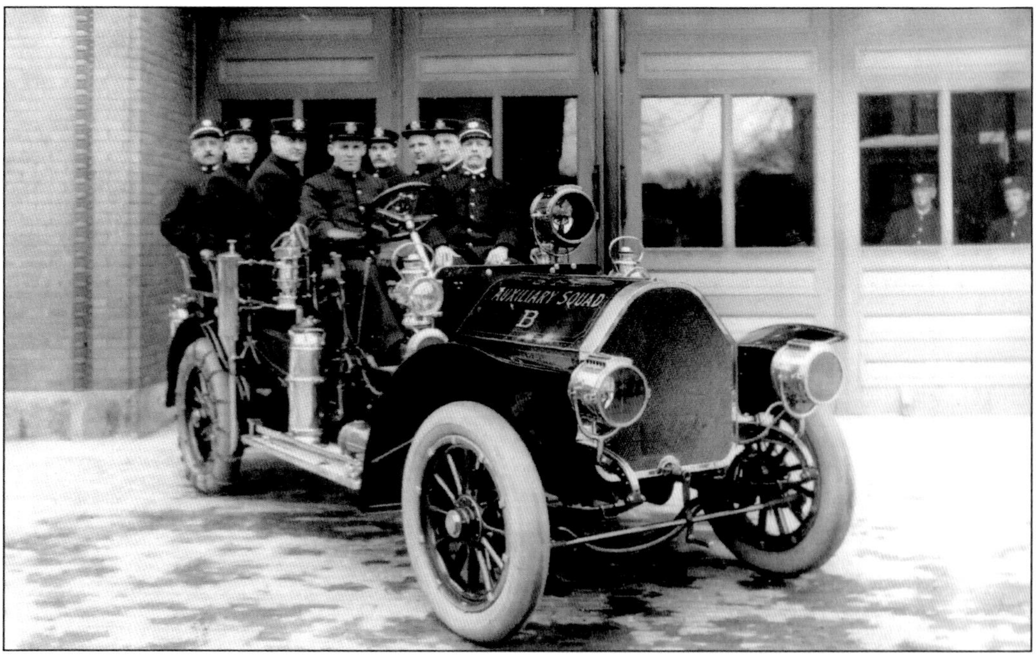

The firefighters and their 1906 Knox hose wagon at Squad B are shown in this 1910 view in front of the Walnut Street fire station. Squad B still runs today as Tactical Unit 1. (Courtesy Noah Pooler family archives.)

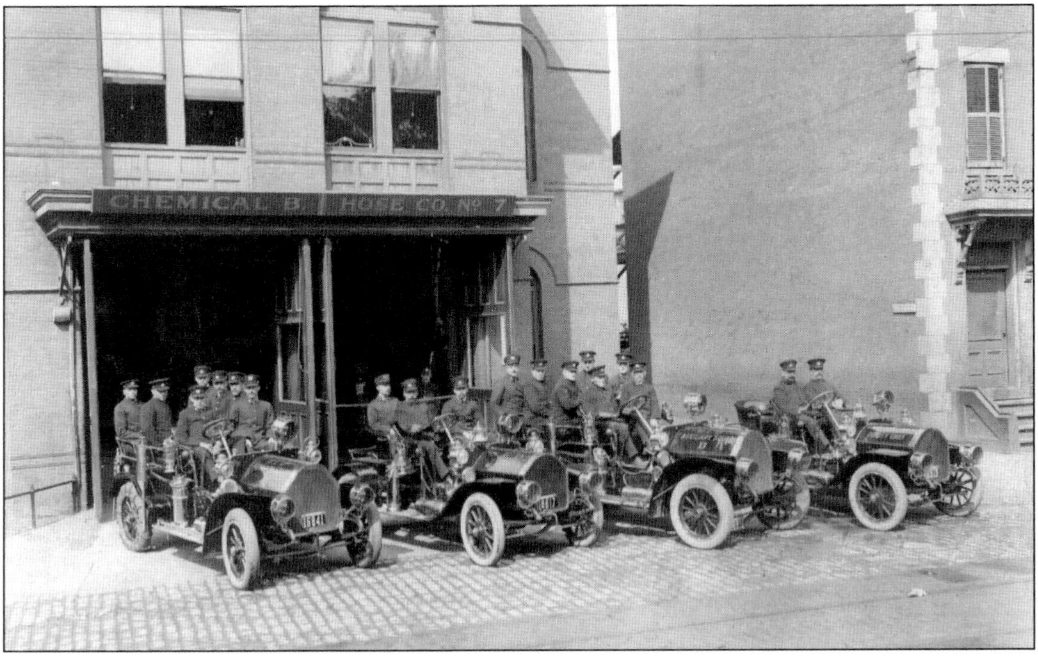

In 1906, Springfield received the first fire apparatus built by the Knox Automobile Company that was located in Springfield. By 1909, the department had five Knox units in operation. This photograph shows two chemical and hose cars operated by the Squads A and B and two cars used by the chief and his assistant. (Courtesy Noah Pooler family archives.)

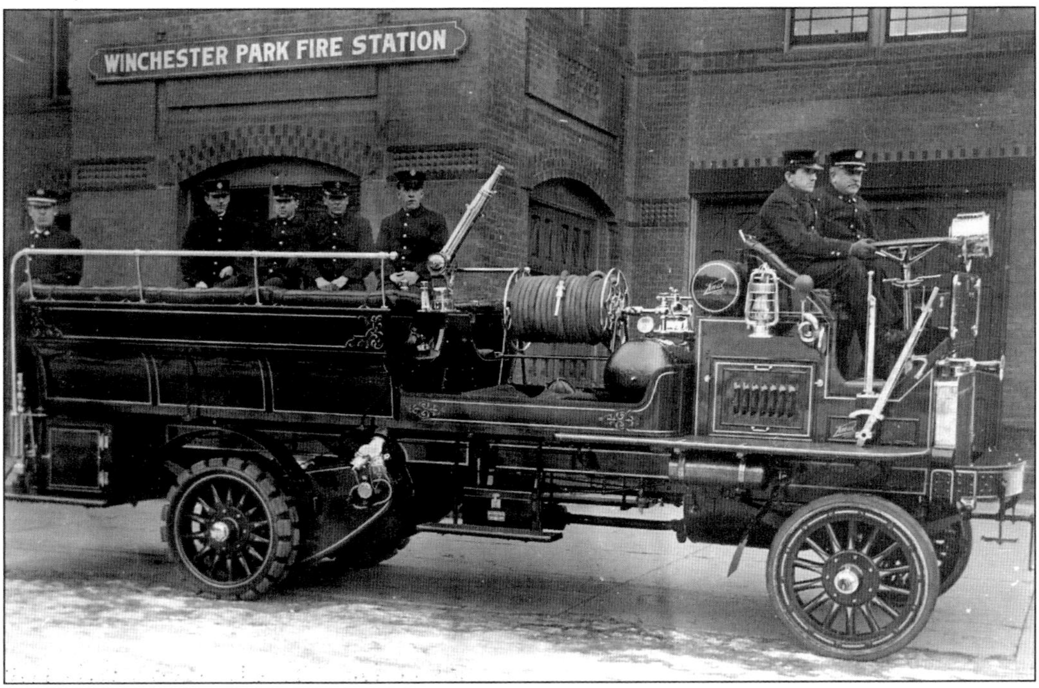

Springfield purchased this Knox chemical and hose wagon in 1907 that served a number of companies in the city over the years. This view shows it running out of the Winchester Park fire station as Hose Company No. 8. (Courtesy Noah Pooler family archives.)

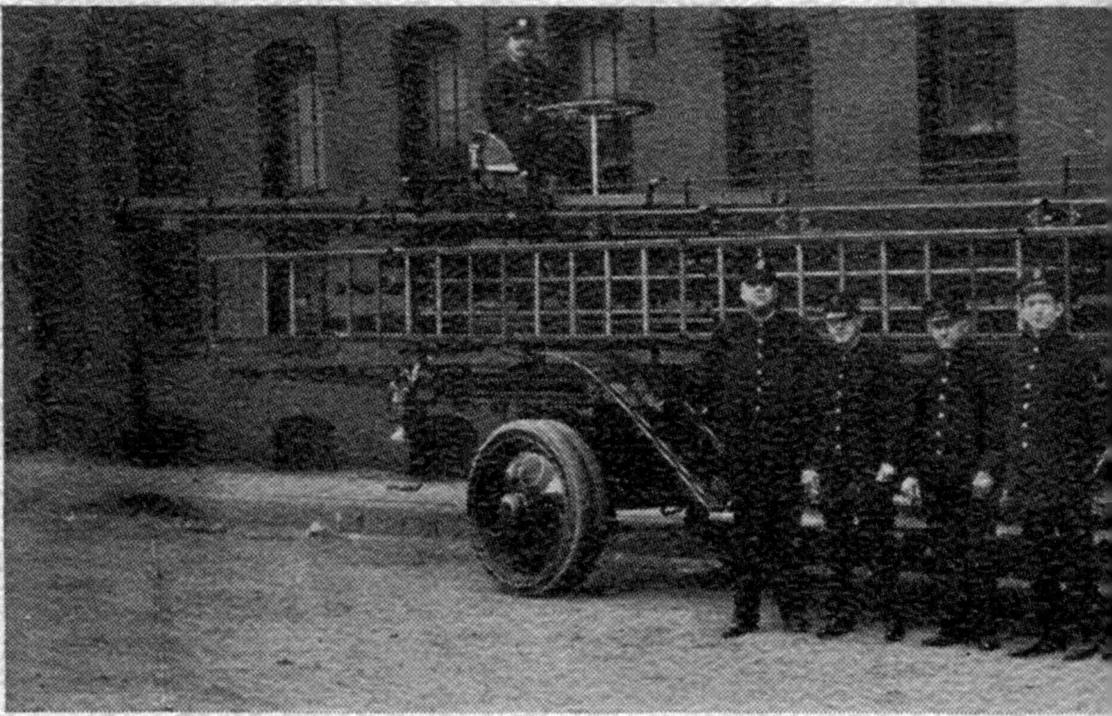

NEW $10,000 MOTOR LADDER TRUCK of Fire

FIRST OF ITS

Weighs over **10** tons, will go **20** miles an hour and fire department by itself, carrying ten different sized rope and a complete equipment of implements for vario

The Seagraves Ladder Company of Columbus, O from the Couple-Gear Freight Wheel Co. of Grand Ra machine, each wheel having a separate motor and ca the power.

In its equipment of motor-fire apparatus SPRI can city, having at present thirteen machines, with been purchased, this being driven by storage electri truck described here.

This advertisement by the Seagrave Company featured the 1910 Seagrave electric-drive ladder truck that operated in Springfield. In the early days of the automobile industry, electric vehicles

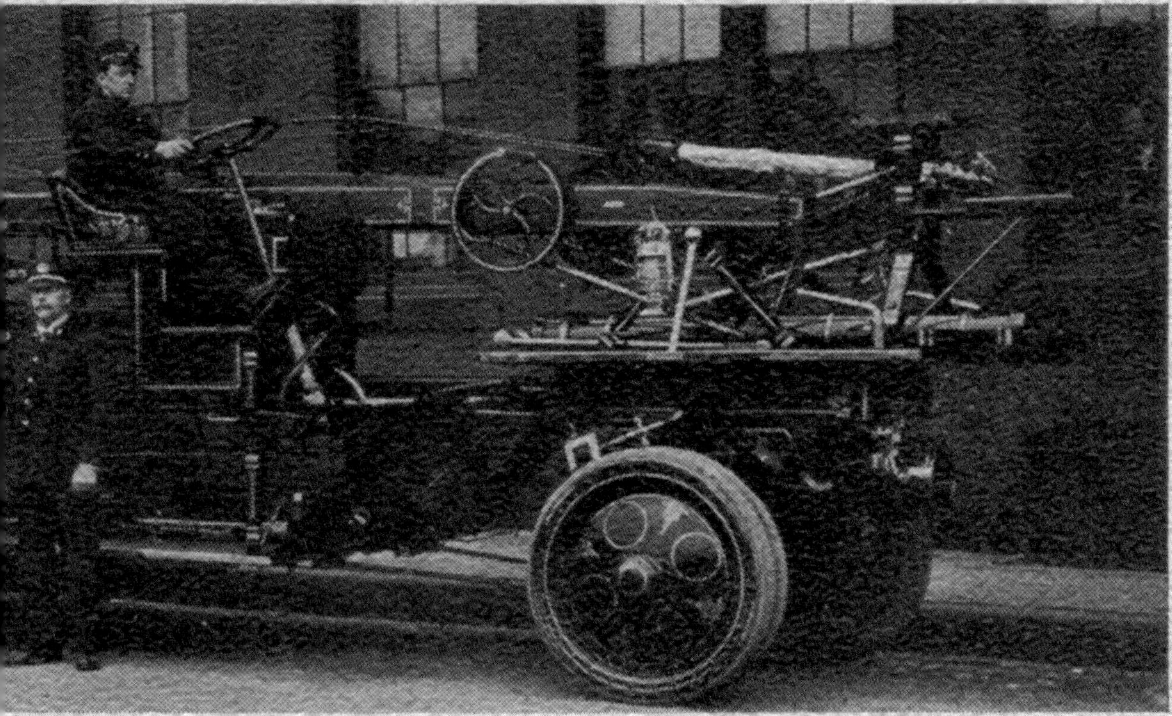

rtment, Springfield, Massachusetts.

D EVER MADE.

b any hill in the city. This truck, **52** feet long, is a small
rs from **18** feet to **85** feet in length, life nets, **150** feet of
es.

nade most of the truck part except chassis, which came
Mich. In these wheels lie the tremendous power of the
operated separately. Electric storage batteries furnish

ELD, MASSACHUSETTS, LEADS every other Ameri-
coming. A combination hose and chemical has lately
tteries and couple-gear wheels, same as the ladder

were still considered superior to primitive gasoline engines, although the batteries available were very inferior compared to the batteries available today. (Courtesy Raymond Pond.)

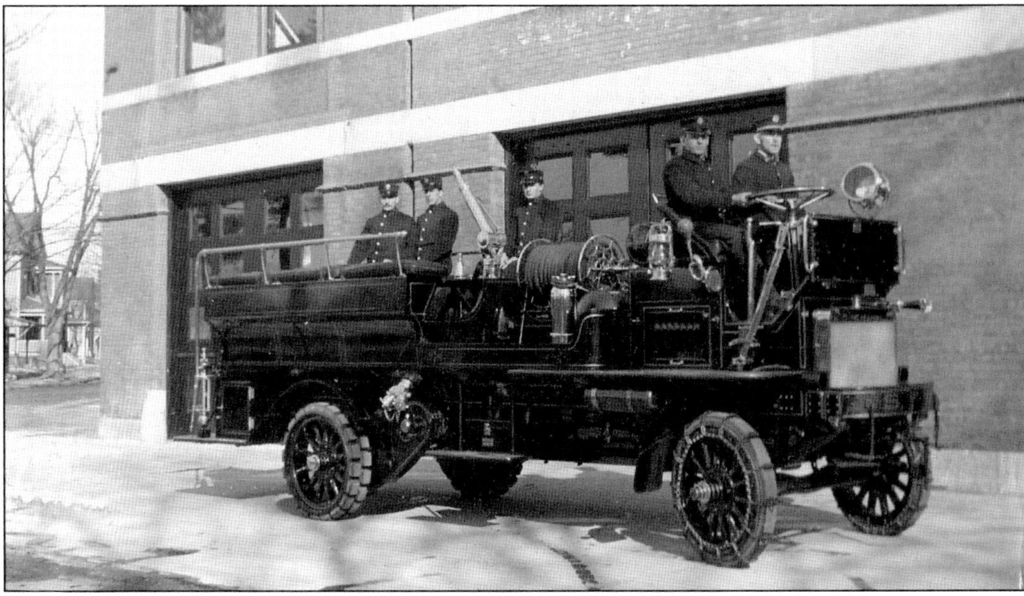

This view shows the 1907 Knox chemical and hose wagon running out of the Pine Street fire station as Chemical Company A. The truck had a 40-horsepower opposed-cylinder engine under the driver's seat. The Combination Ladder Company of Providence, Rhode Island, built the hose body for Knox. The maximum speed was 20 miles per hour with a full load of hose and five firefighters. The driver's seat above the engine is a carryover from the horse-drawn days. (Courtesy Noah Pooler family archives.)

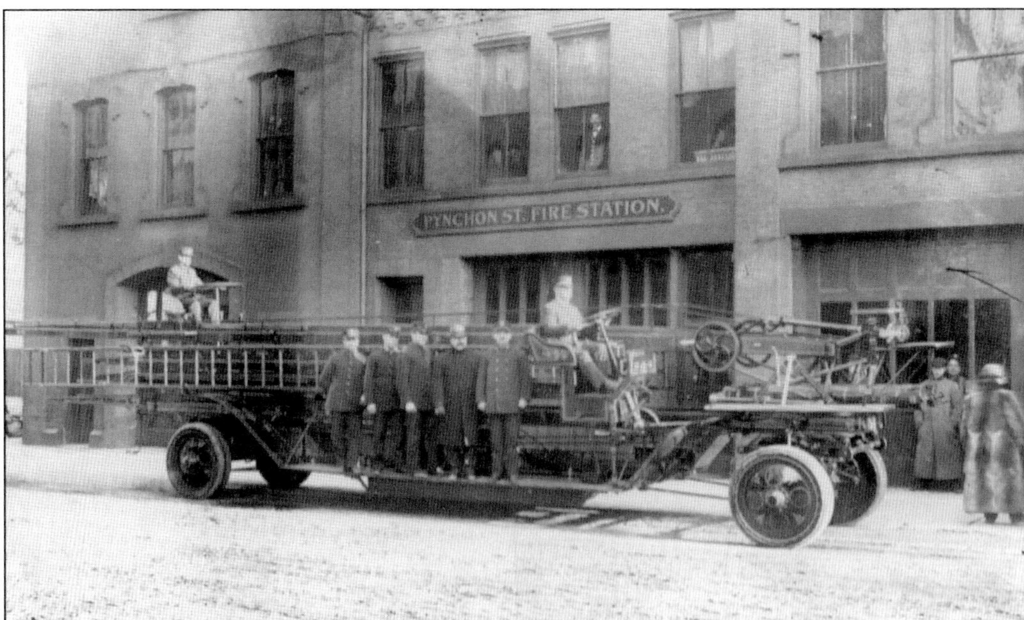

The 1910 Seagrave 85-foot aerial ladder is shown in front of the Pynchon Street fire headquarters in 1911. (Courtesy Noah Pooler family archives.)

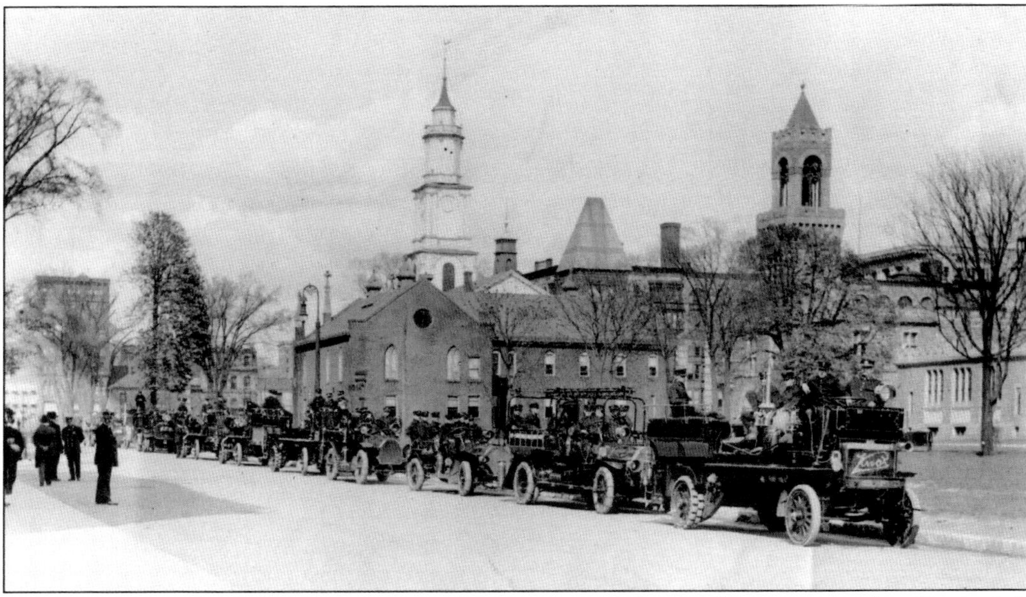

This view from about 1914 shows some of the 24 pieces of Knox fire apparatus that Springfield operated. The Springfield Fire Department was a good customer to the Knox Automobile Company, which built all its fire apparatus in Springfield. Most people today think that American LaFrance or Seagrave were pioneers in motor fire apparatus, but Knox was the first to introduce one in 1906. Seagrave did not introduce its first successful truck until 1907 and American LaFrance not until 1910. (Courtesy Raymond Pond.)

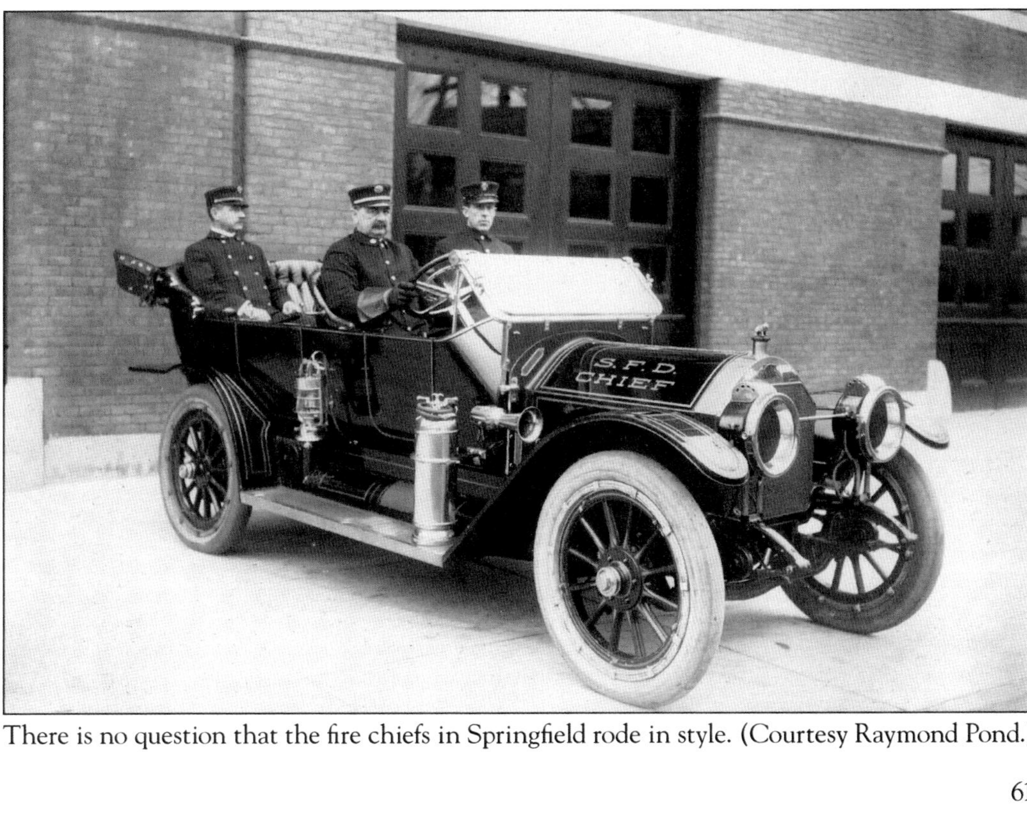

There is no question that the fire chiefs in Springfield rode in style. (Courtesy Raymond Pond.)

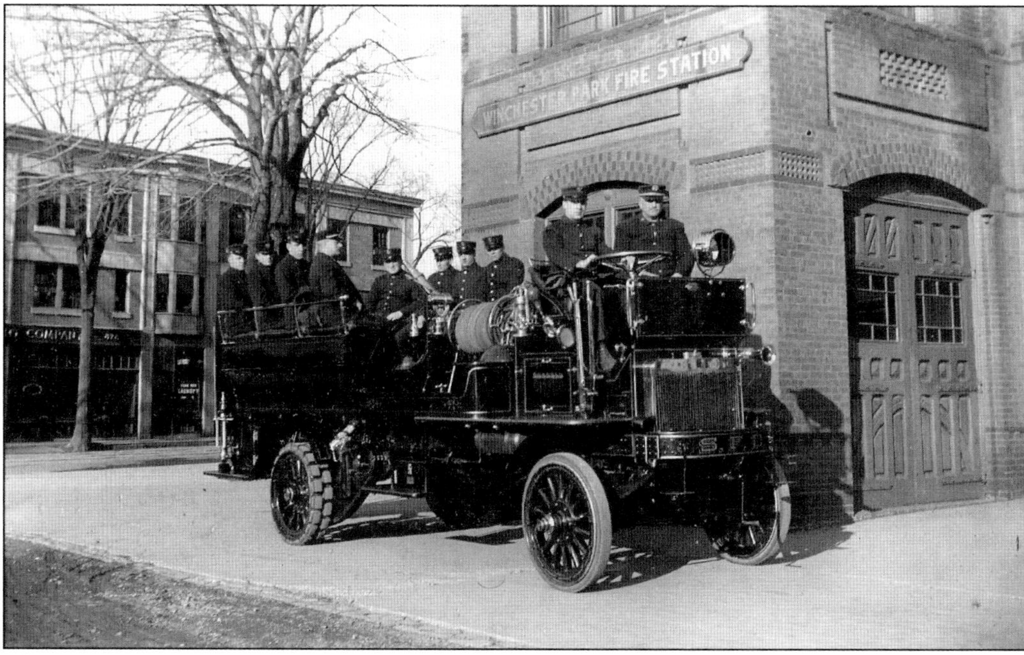

This is another view of Springfield's 1907 Knox chemical and hose wagon in front of the Winchester Square fire station. Many departments all over the United States operated similar units; the fire department in Hartford, Connecticut, had an identical unit that was in service until 1937. (Courtesy Noah Pooler family archives.)

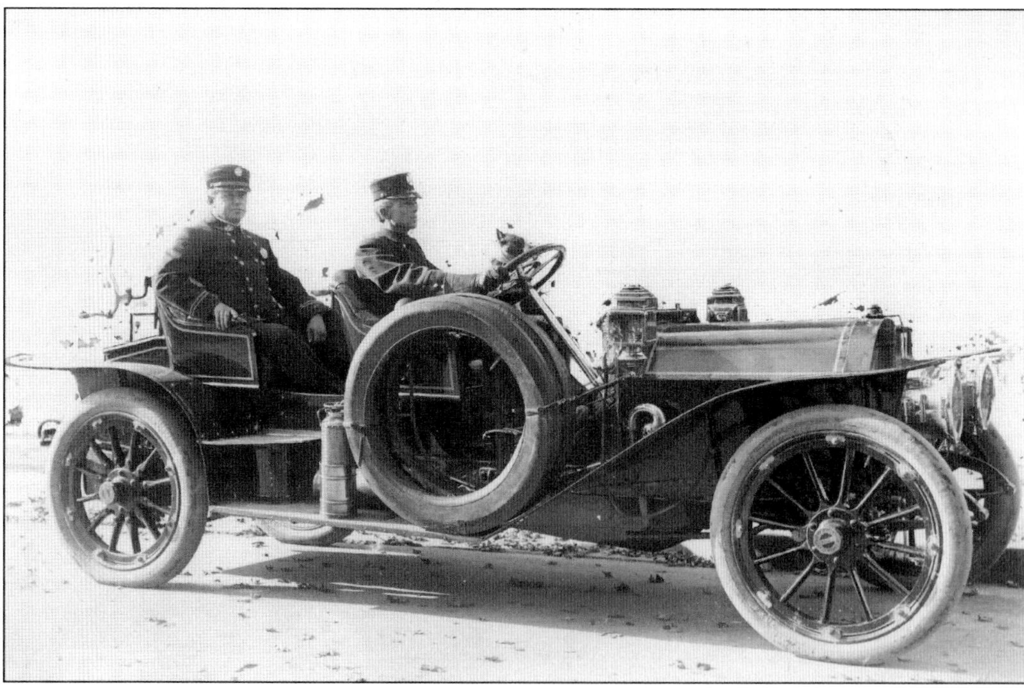

This view from around 1914 shows the automobile used by fire alarm superintendent Samuel Wheeler. The make of this car is unknown, although it may be a Locomobile that was built in Bridgeport, Connecticut, before World War I. (Courtesy Connecticut Fire Museum archives.)

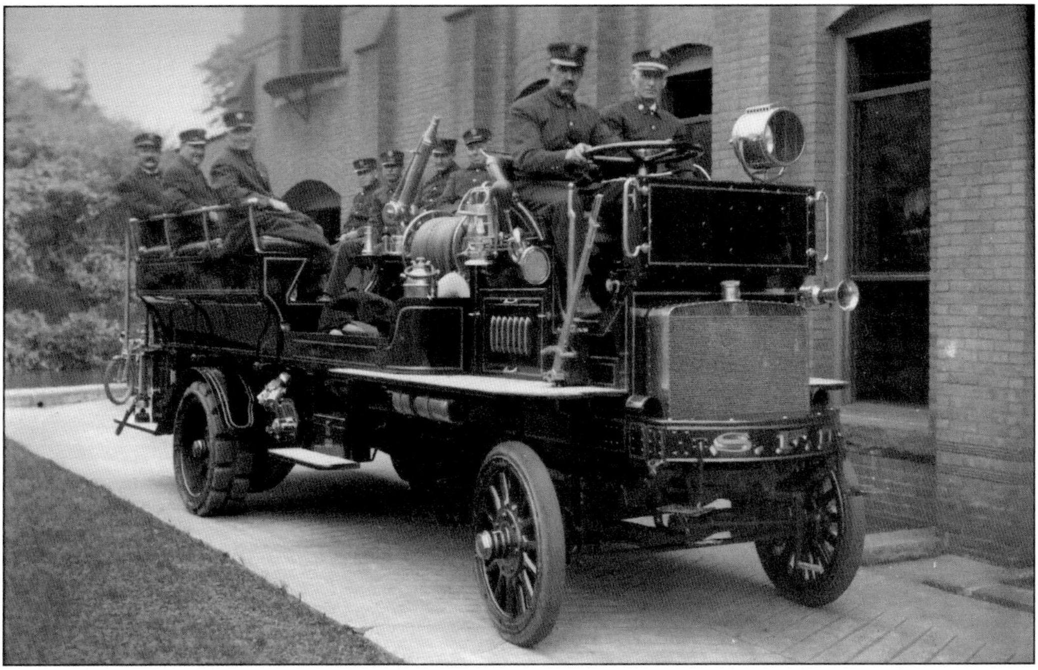

This view shows the 1907 Knox chemical and hose wagon next to the Mill Street fire station in 1910. This was before red warning lights were put on fire apparatus. It was not until the 1930s that fire apparatus or police cars had warning lights. (Courtesy Noah Pooler family archives.)

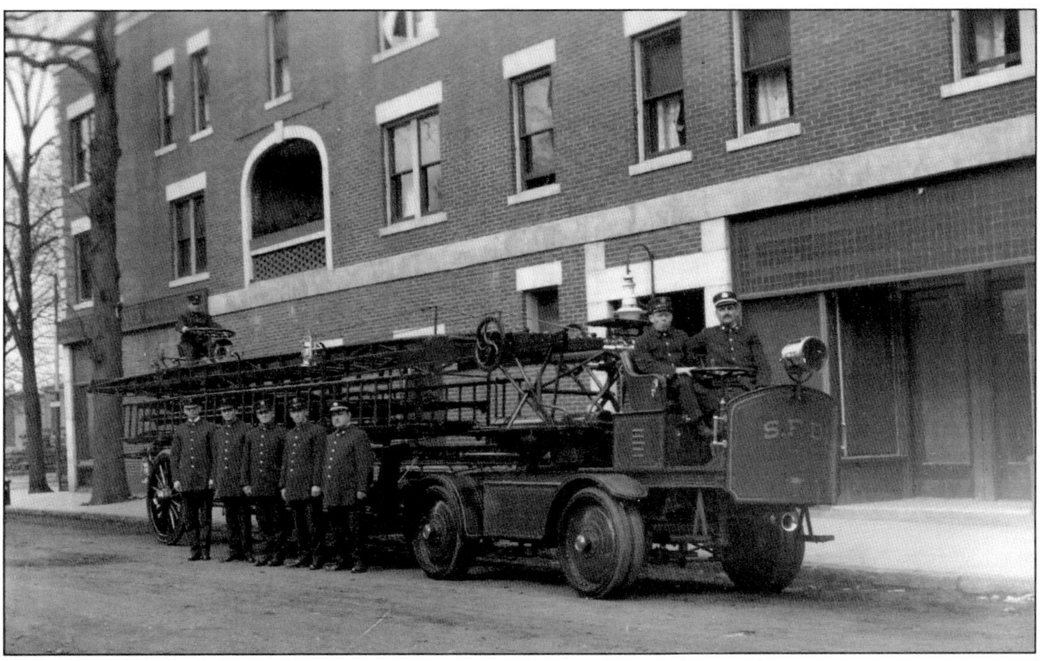

The Couple Gear Freight Wheel Company of Grand Rapids, Michigan, was a purveyor of front-wheel-drive and all-wheel-drive trucks. This one is attached to 1914 Seagrave 85-foot aerial ladder truck. Couple Gear used a gasoline engine to drive a generator that supplied electricity to the four electric motors, one on each wheel. (Courtesy Noah Pooler family archives.)

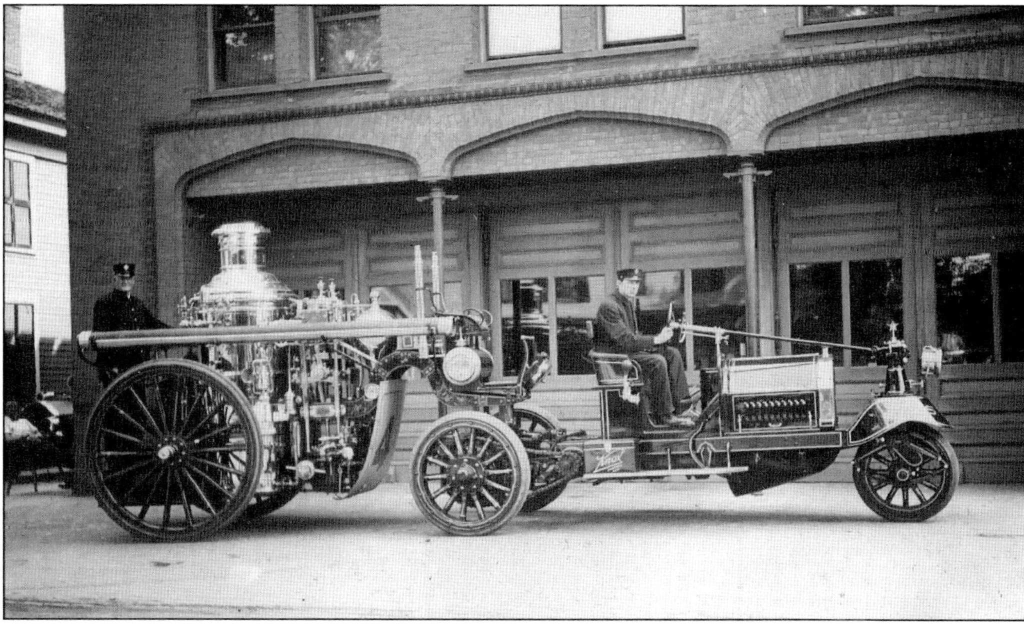

This photograph shows a 1903 American LaFrance 700-gallon-per-minute steamer with a Knox-Martin tractor from Engine Company No. 4. The gentleman on the back of the steamer is Mr. Burton, who was 72 years old when this picture was taken in 1914. He was the oldest call firefighter on the department. Call firemen were paid but only reported to the firehouse when an alarm came in. (Courtesy Noah Pooler family archives.)

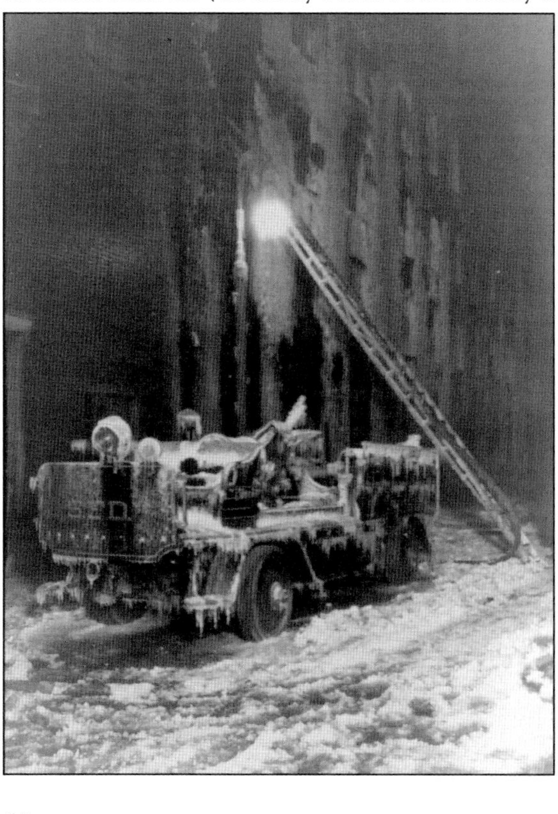

This is a winter view of Springfield's 1914 Seagrave-Couple Gear chemical and hose wagon at a fire on January 18, 1921, at the Hinsdale Smith Company located at 23 Hampton Street. Fires that occurred during the winter months caused considerable hardship for the firefighters and their equipment. (Courtesy Connecticut Fire Museum archives.)

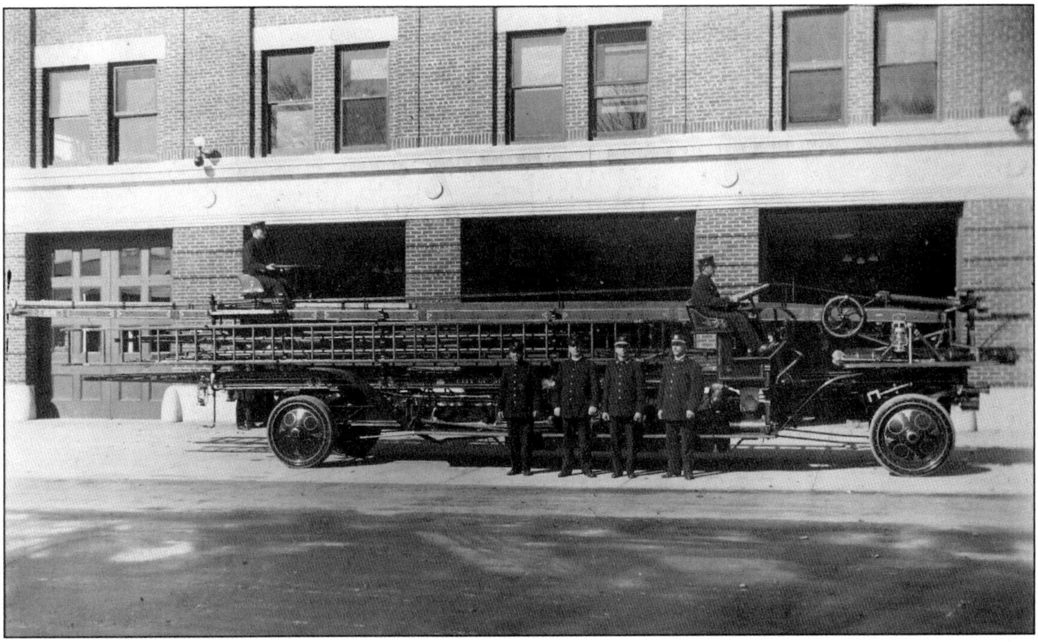

The 1910 Seagrave 85-foot aerial ladder truck is shown in this 1920s view in front of the new fire headquarters at Court Street and Columbus Avenue. (Courtesy Noah Pooler family archives.)

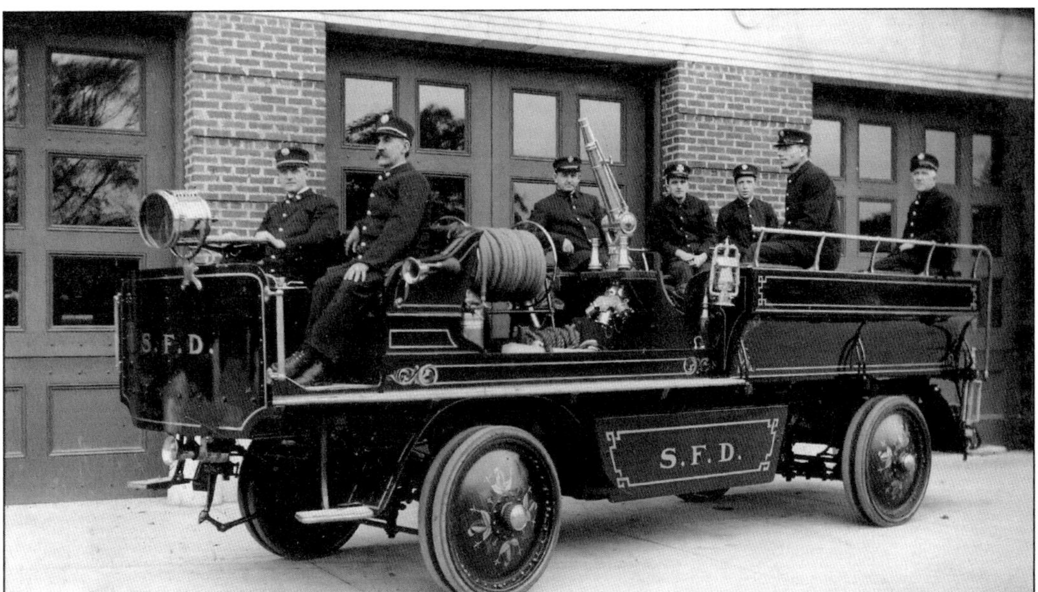

In 1914, Seagrave delivered this combination chemical and hose wagon that was built on a Couple Gear chassis. Springfield pioneered the squad body that allowed all the firefighters to ride inside the apparatus. Most departments still had the firemen riding on the rear step of the pumper and on the side of the ladder trucks. Over the years, many firemen were killed or severely injured when they fell off the truck responding to a fire alarm. Today all firefighters are required to have a seat on all apparatus and no one is allowed to ride on the rear step. (Courtesy Noah Pooler family archives.)

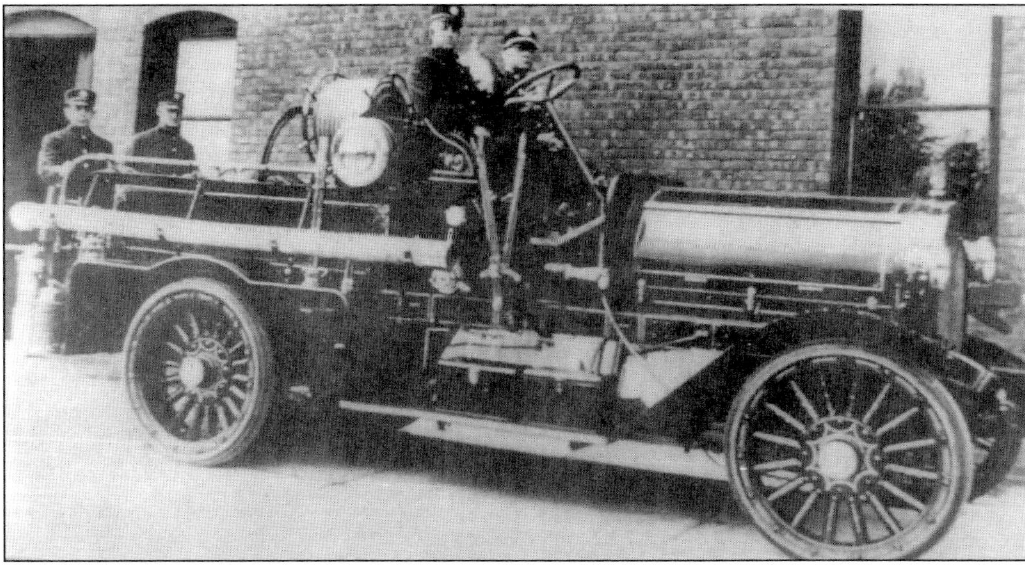

The W. S. Nott Company of Minneapolis, Minnesota, built this combination motor pumper and hose car for Springfield in 1913. Nott had a 600-gallon-per-minute piston pump similar to the pumps the company used in its steam pumpers. It was not successful in making the transition from steam to motor fire apparatus and was out of business by 1916. Two similar engines were also delivered to New York City in 1913. (Courtesy Connecticut Fire Museum archives.)

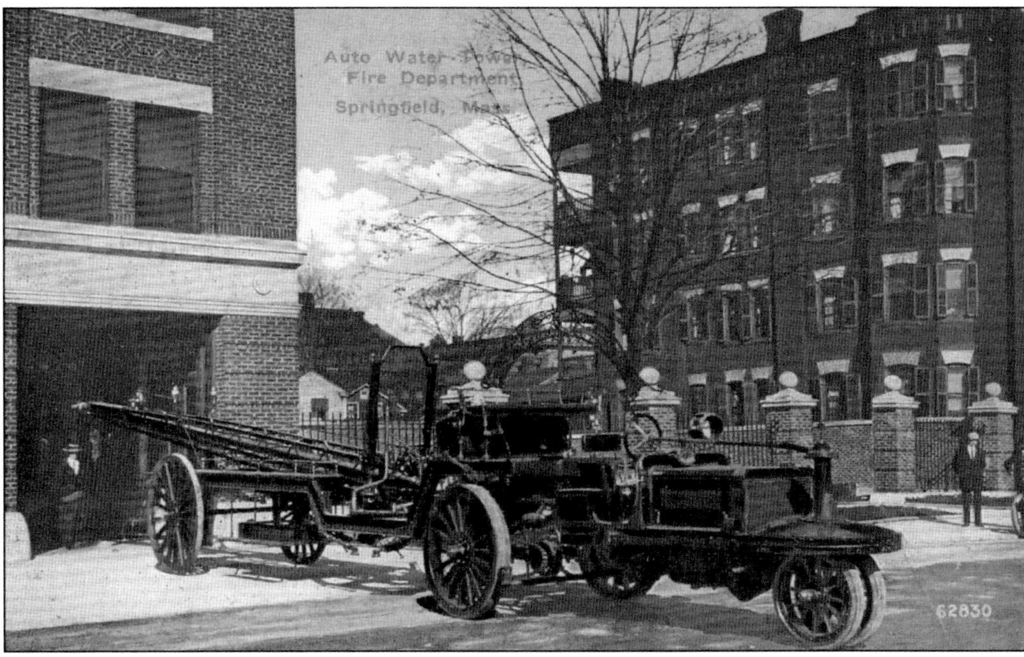

An early Knox-Martin tractor was placed under the 1894 Champion water tower in 1912 in an effort to modernize the tower. Note the friction brakes on the rear wheels similar to a wagon. The Knox-Martin tractors were designed by Charles H. Martin, who invented the rocking fifth-wheel trailer hitch still used today. The three-wheeled tractor was designed to motorize horse-drawn wagons. These were only built from 1912 to 1915, when a more conventional tractor was designed. (Courtesy Geoffery Neilsen.)

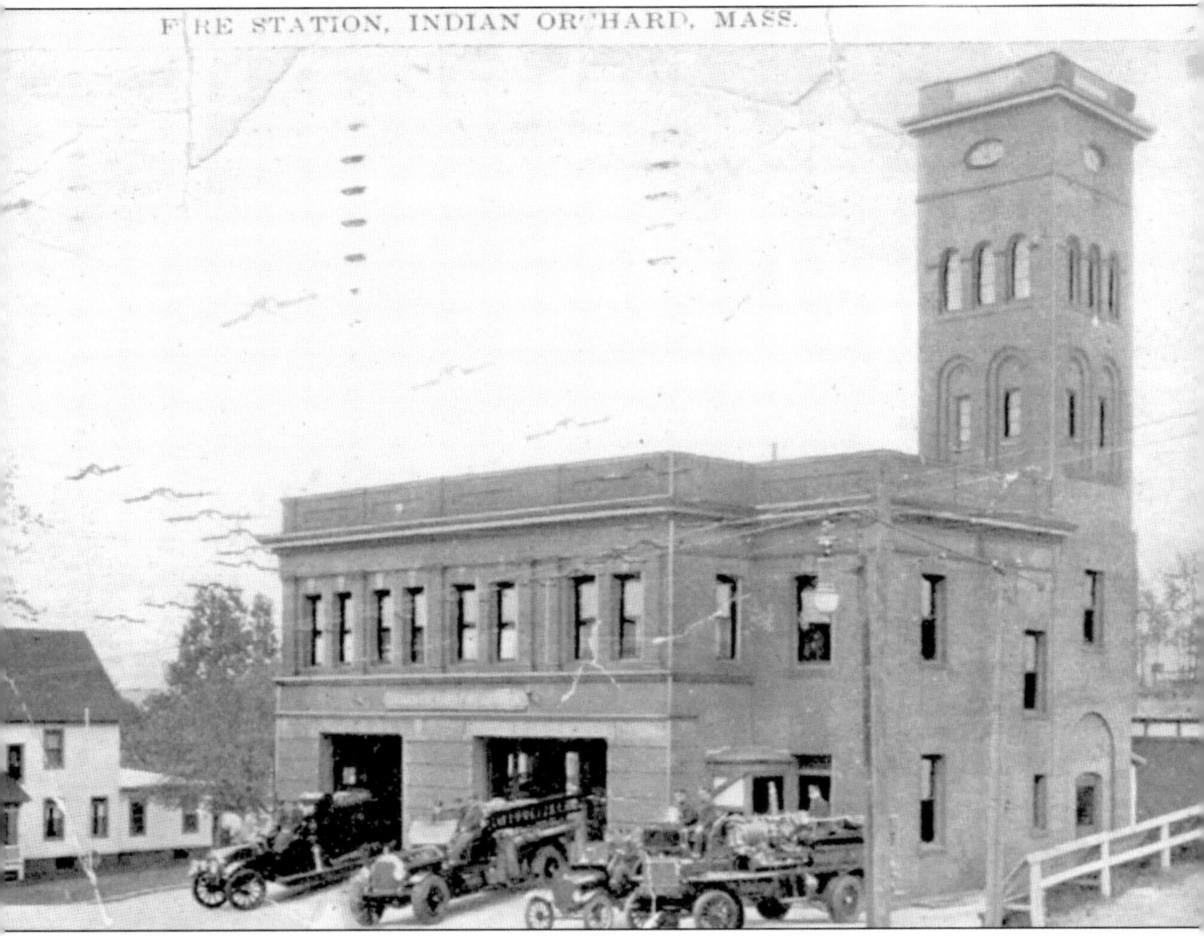

This is a view of Indian Orchard fire station in 1920 with its apparatus, including a 1907 Knox chemical and hose wagon, two Seagrave pumpers, and a deputy chief's run-a-bout car. Fire protection in Indian Orchard began in 1861, when Champion Fire Company was organized and equipped with a hand-drawn pumper and hose cart. (Courtesy Geoffery Neilsen.)

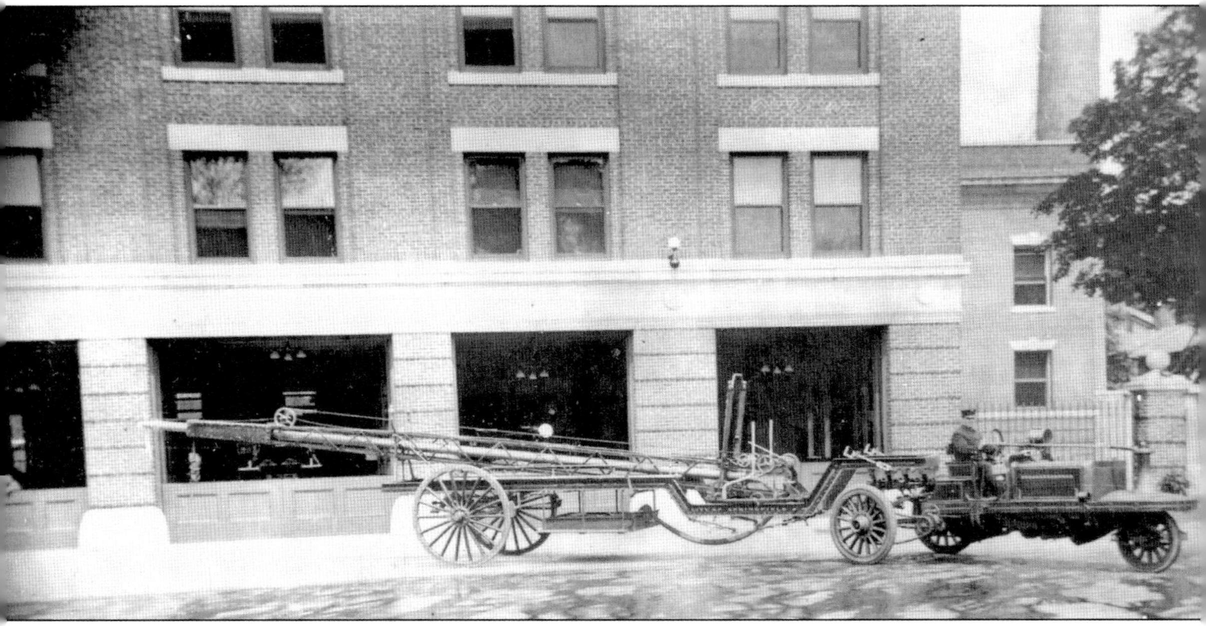

The water tower with a Knox-Martin tractor is shown in front of the Court Street headquarters. Note the Dutch doors that allowed fresh air into the firehouse but still kept the horses inside if they were out of their stalls. (Courtesy Noah Pooler family archives.)

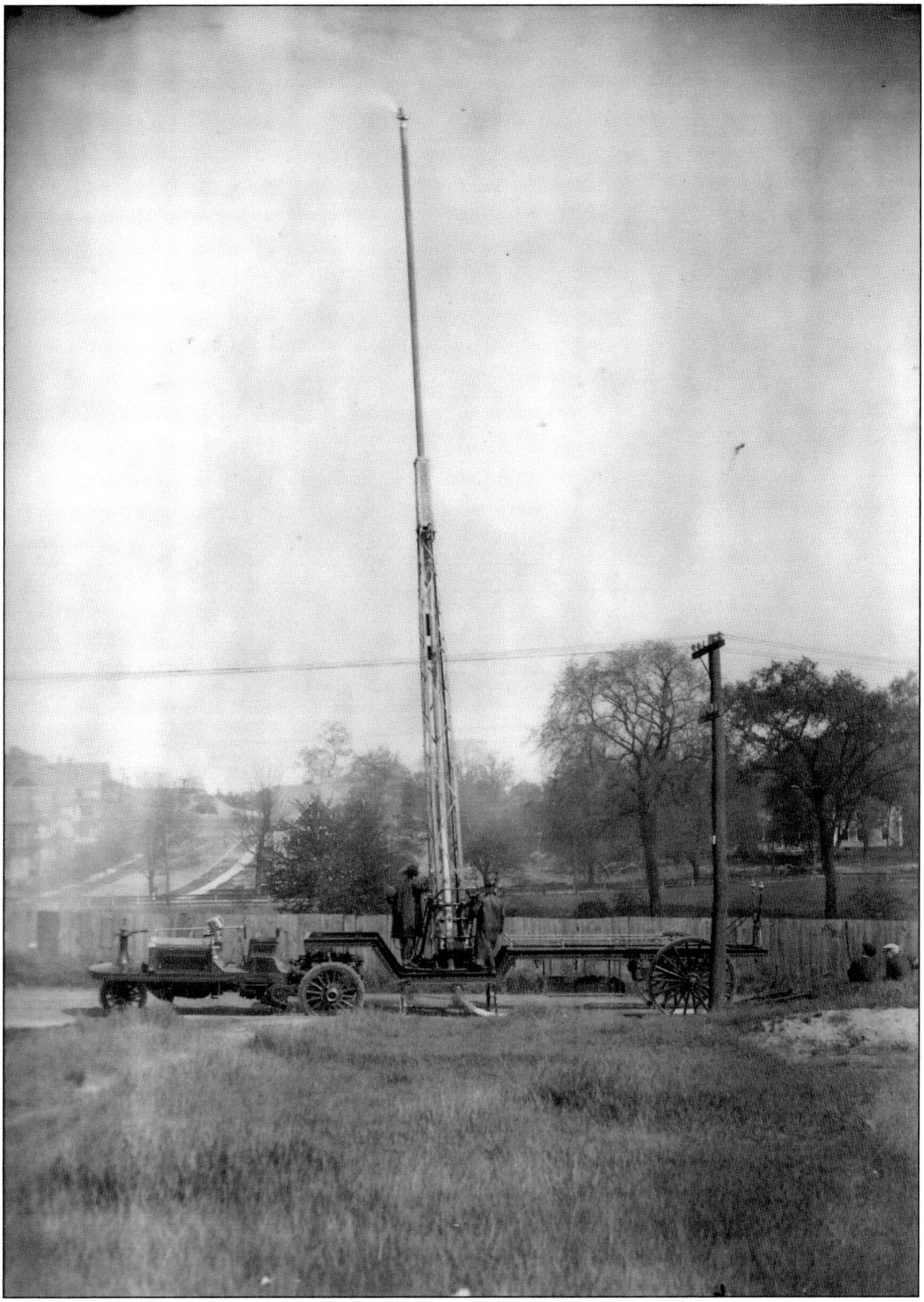

The Champion water tower is shown in this 1914 view of a drill at the Watershops pond of the Springfield Armory. (Courtesy Noah Pooler family archives.)

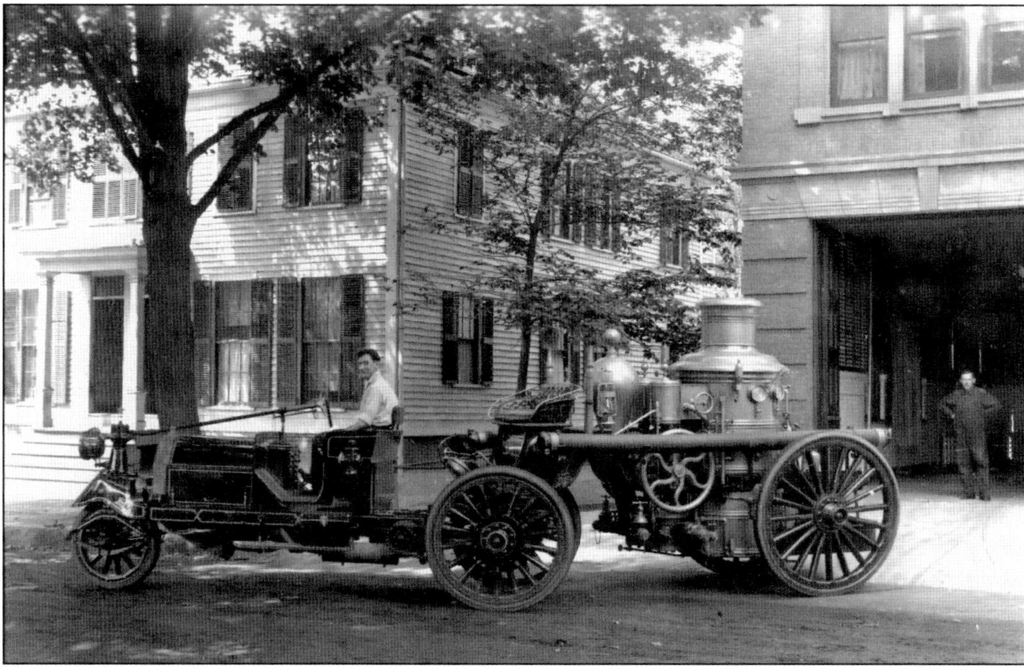

This is a side view of a Knox-Martin tractor pulling one of Springfield's steams pumpers. By 1920, all the steam pumpers were retired. (Courtesy Noah Pooler family archives.)

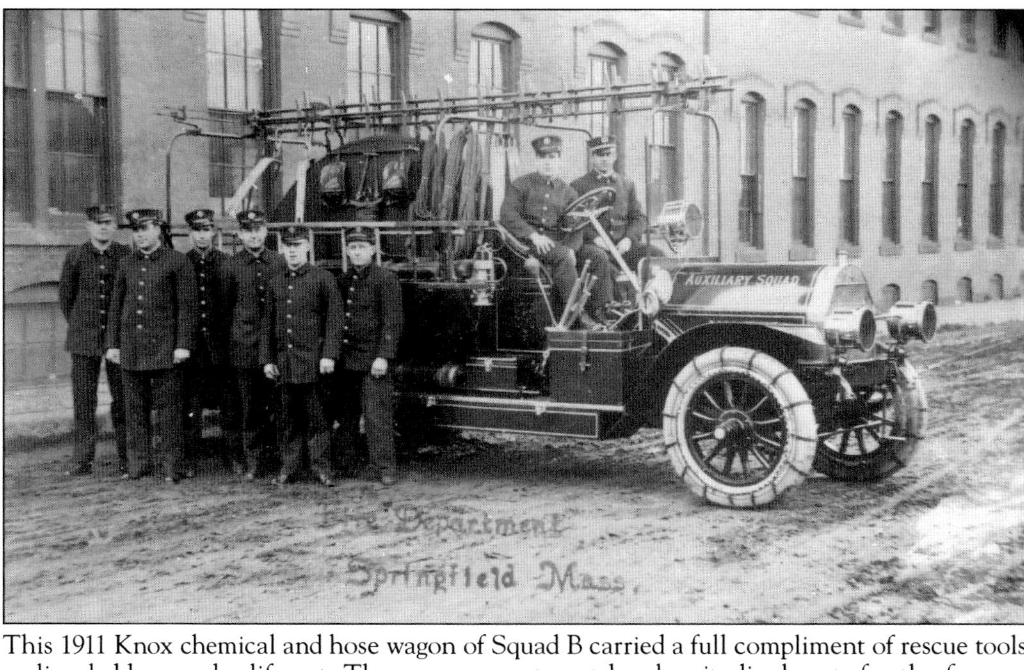

This 1911 Knox chemical and hose wagon of Squad B carried a full compliment of rescue tools, scaling ladders, and a life net. The rear compartment has longitudinal seats for the firemen. An acetylene tank was also mounted on the truck that powered the headlights and spotlight. (Courtesy Geoffery Neilsen.)

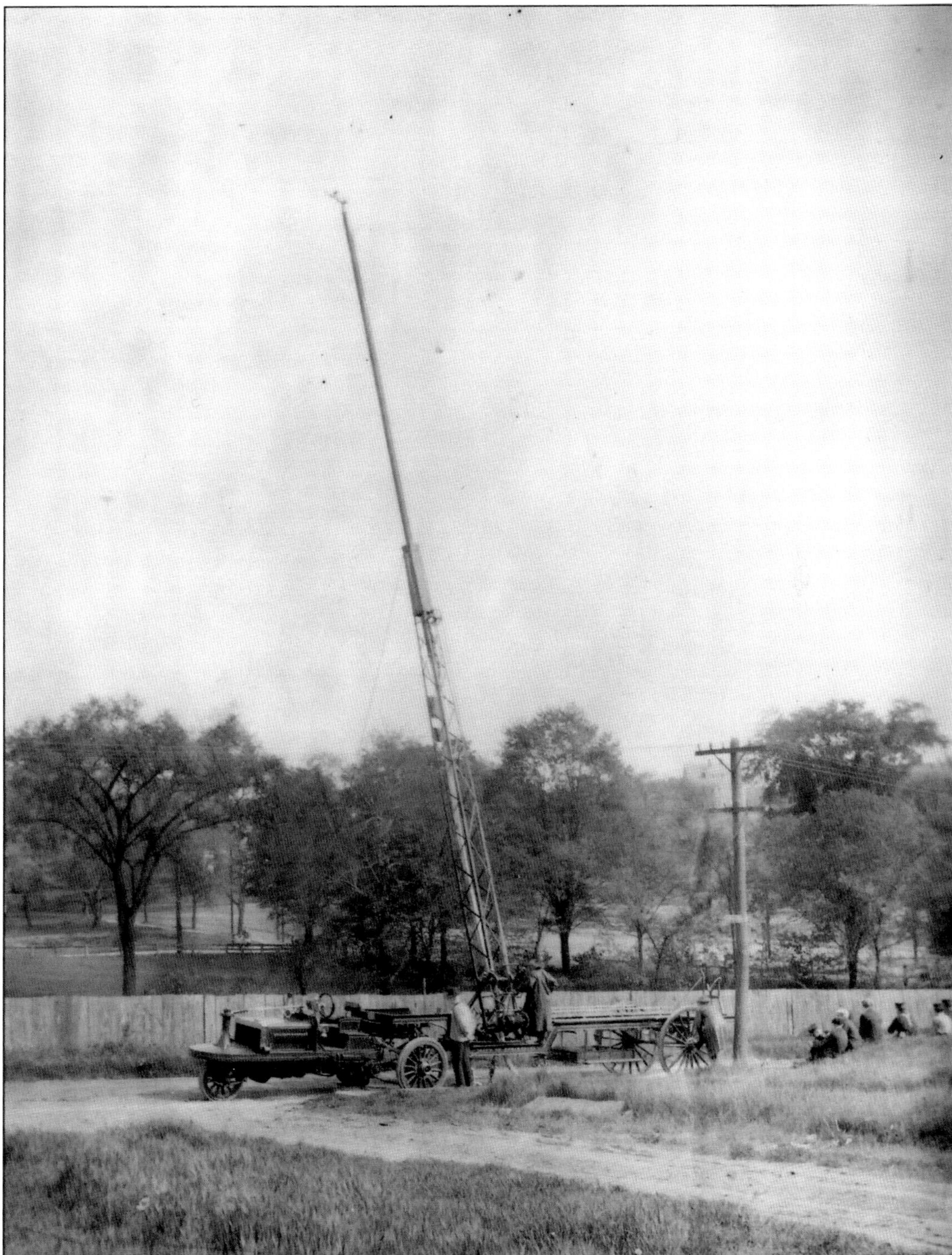

The water tower was a specialized piece of equipment designed to put heavy streams of water into the upper floors of burning buildings. In March 1955, the Springfield water tower was sold to Charles F. Hawkins of Willimantic, Connecticut, for $301.50, where it was subsequently scrapped. (Courtesy Noah Pooler family archives.)

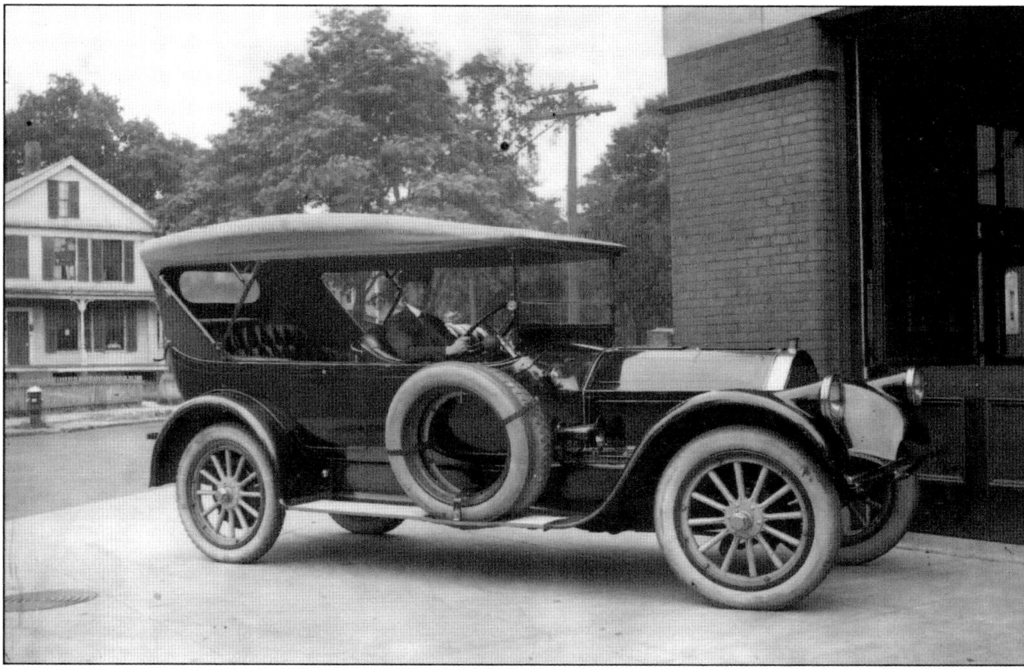

The department used this Pierce-Arrow automobile during the 1920s for one of the assistant chiefs. (Courtesy Raymond Pond.)

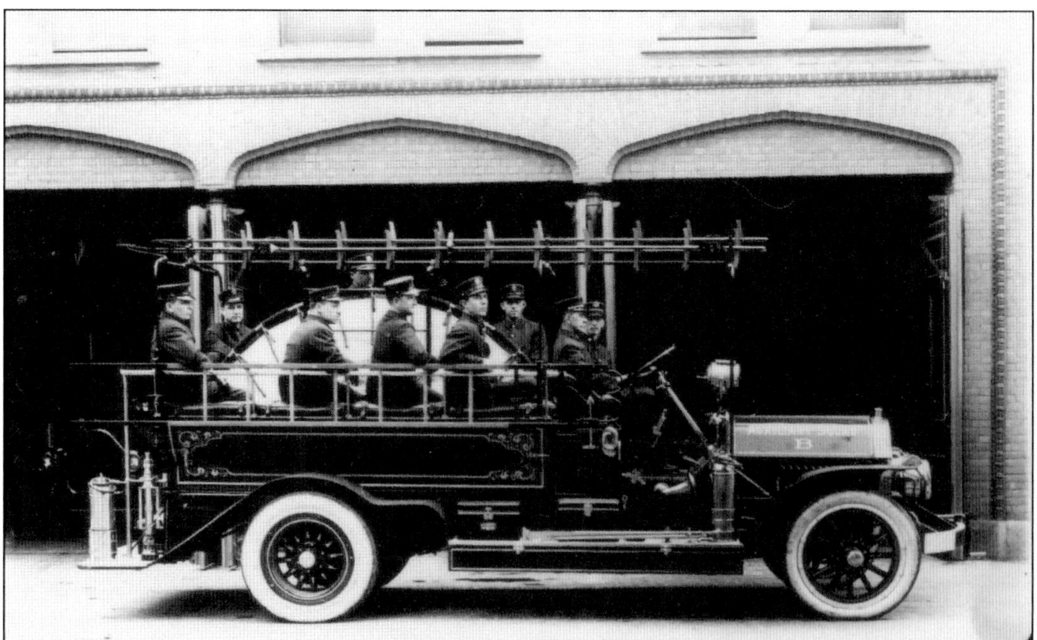

This view of Squad B and its 1911 Knox chemical and hose wagon shows how the firefighters rode to the fire on the seats along the inside of the hose bed. (Courtesy Noah Pooler family archives.)

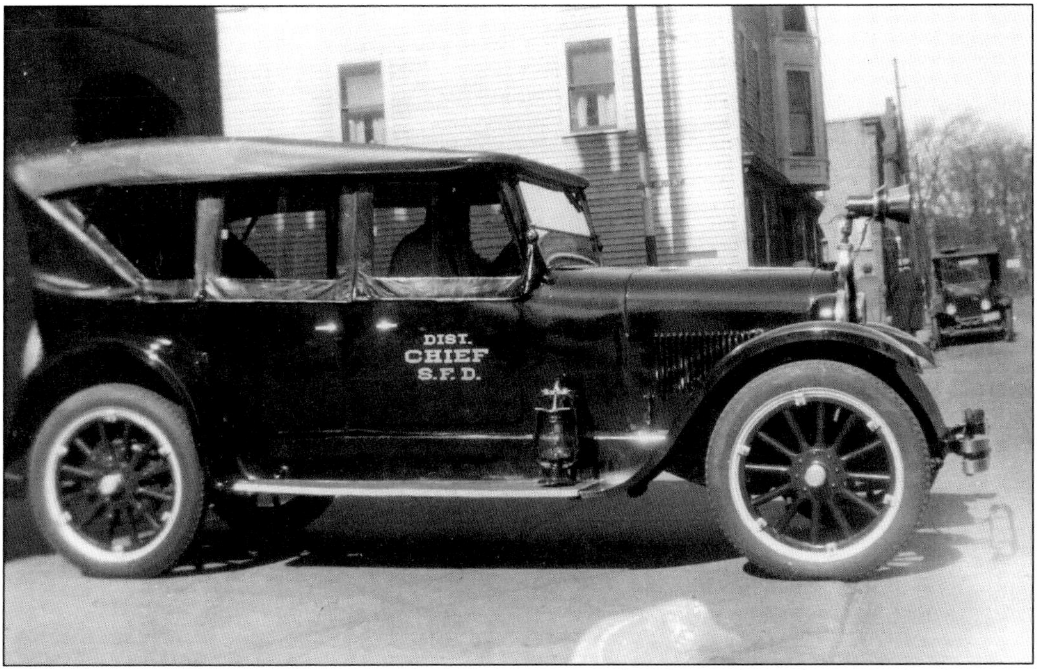

This view shows another district chief's car. This one looks like a Chevrolet or Dodge from the early 1920s. (Courtesy Raymond Pond.)

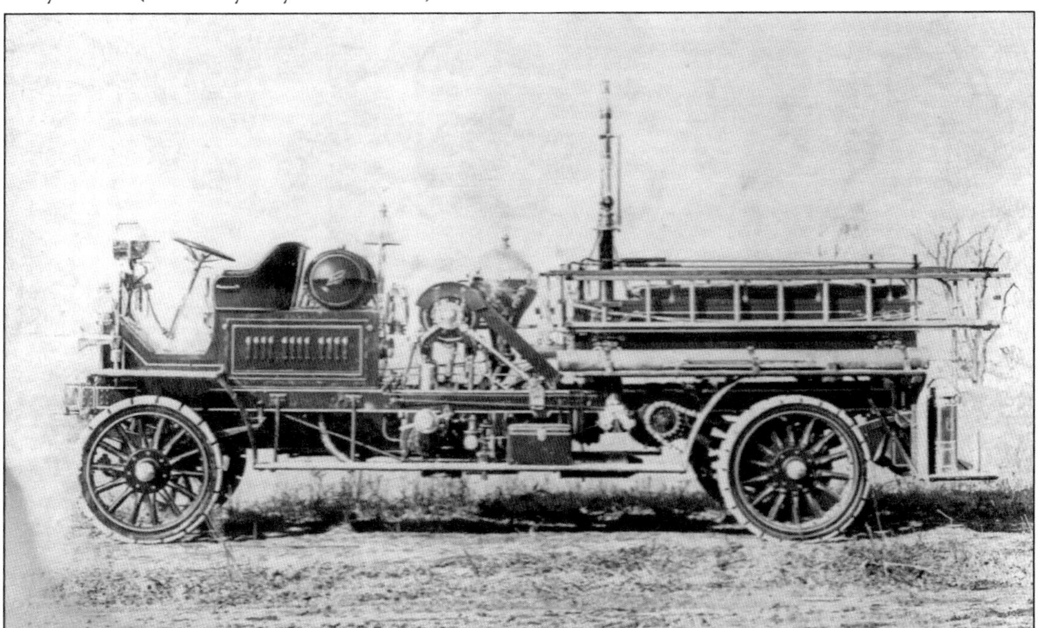

The Knox Automobile Company was founded in 1898 in Springfield and was one of the first to develop and build a water-cooled automobile. In 1906, the company began to build fire apparatus and provided some of the first motorized apparatus to fire departments all over the country. This builder's photograph shows a 1913 Knox pumper at the factory before delivery to the Springfield Fire Department. The pump on this truck was a piston pump of the Amoskeag type similar to those used on steam fire apparatus. (Courtesy Connecticut Fire Museum archives.)

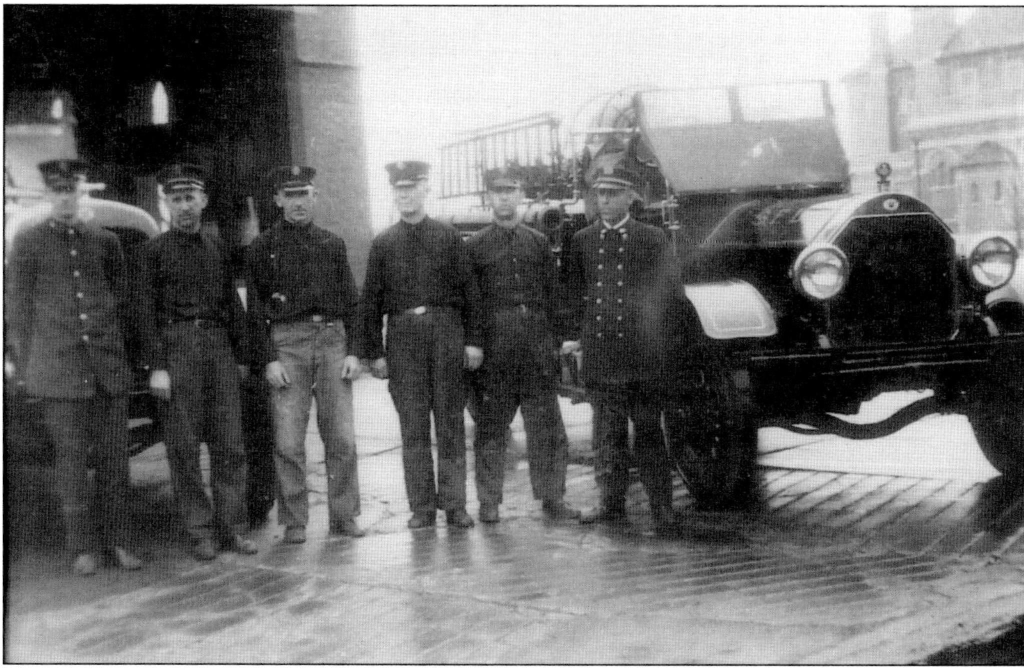

This view shows both 1916 and 1930 Seagrave pumpers at the Oakland Street fire station around 1930. (Courtesy Geoffery Neilsen.)

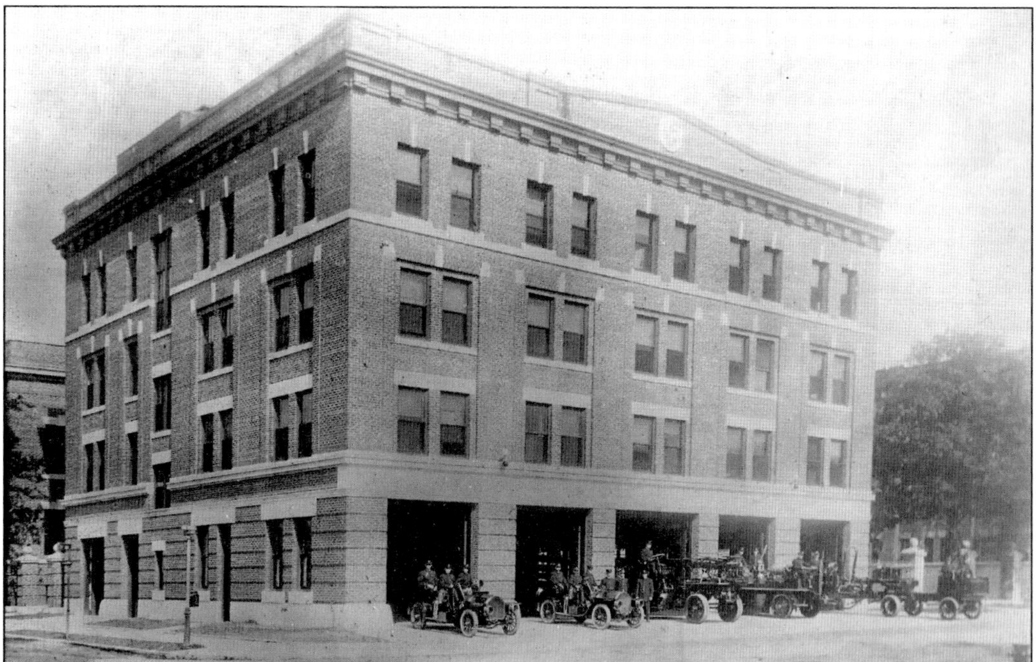

This view from around 1920 shows the Court Street fire headquarters along with some of the apparatus that operated out of that station. By this time the Knox-Martin tractor on the water tower had been replaced by a rebuilt 1907 Knox hose wagon. (Courtesy Connecticut Fire Museum archives.)

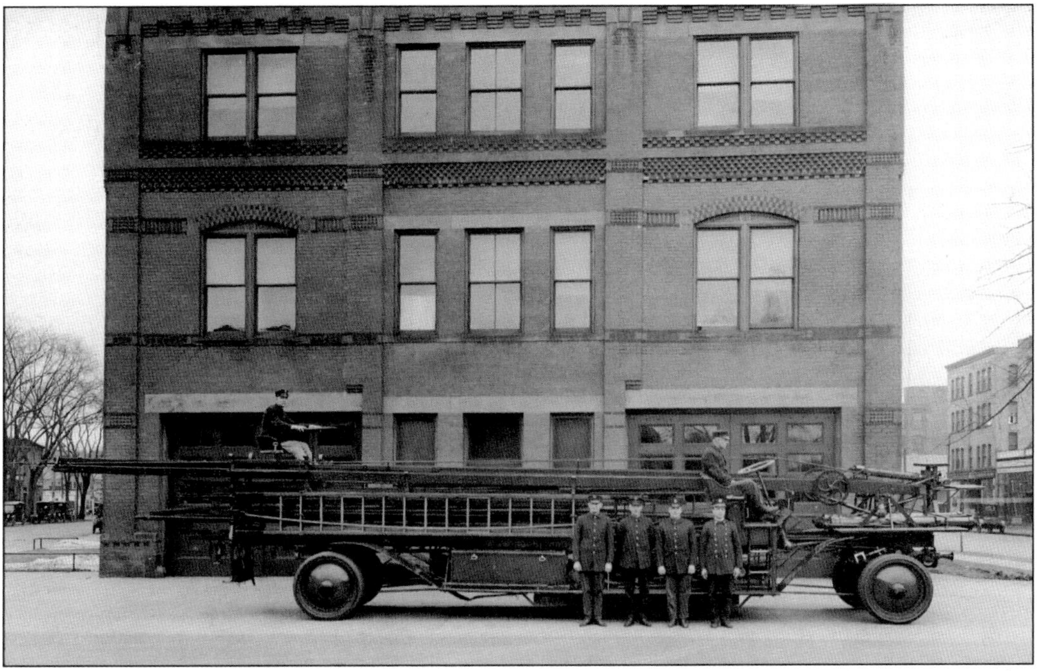

This view shows the 1910 Seagrave 85-foot aerial ladder in front of the Winchester Square fire station after the station was remodeled in 1929. This truck operated as Ladder Company No. 8 from this station. (Courtesy Noah Pooler family archives.)

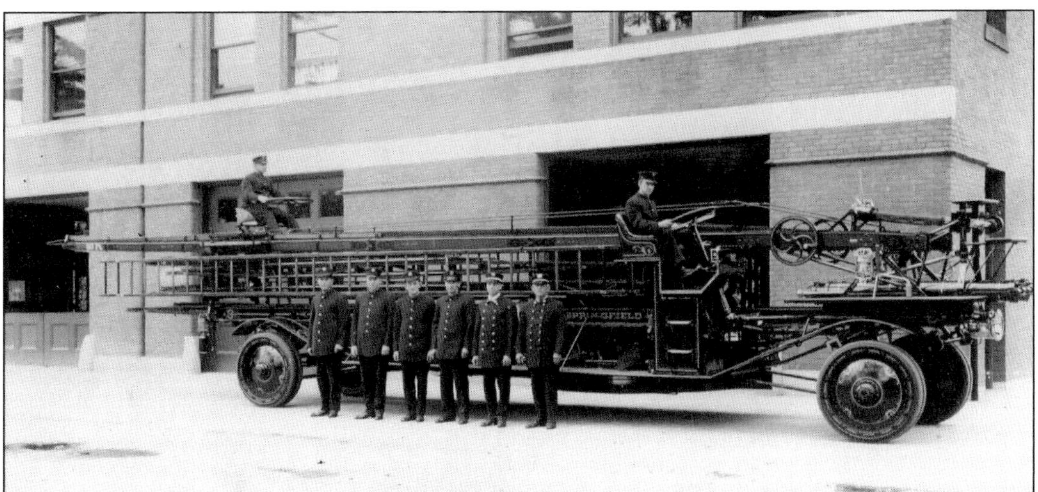

The 1910 Seagrave 85-foot aerial ladder truck had an electric drive that used batteries to drive motors on each wheel. The truck was not necessarily designed to be operated in a city such as Springfield. It was underpowered and usually exhausted its batteries running to the fire, which necessitated towing it back to the station to have its batteries recharged after the fire. Instead of sitting up front, the driver sat well back on the right-hand side of the vehicle. This design did not give the driver a good view of where he was going. This truck was converted to a tractor-drawn ladder truck with an Autocar tractor in 1934. After being purchased by a collector in the 1960s, the truck sat for many years derelict in a field in Cummington. (Courtesy Noah Pooler family archives.)

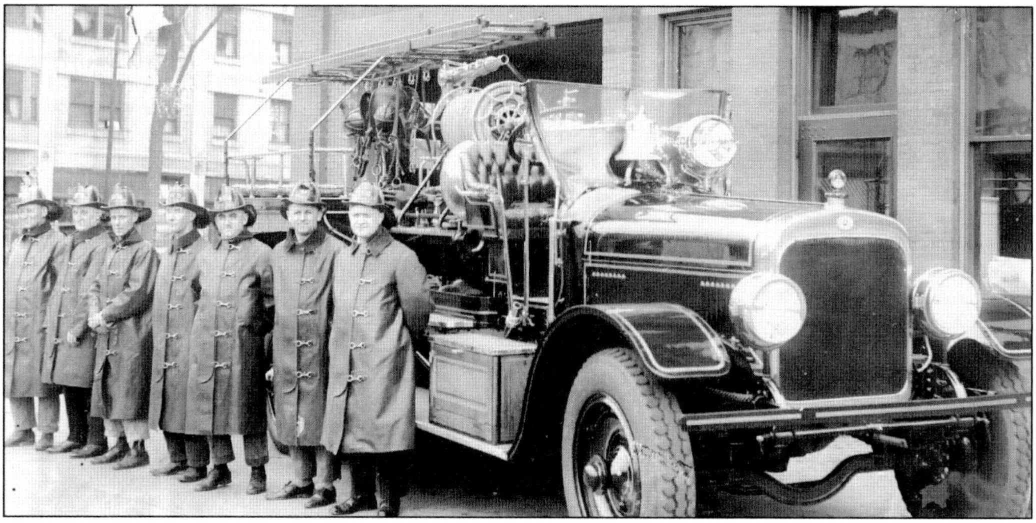

One of the 1930 Seagrave 1,000-gallon-per-minute pumpers is shown with its firemen in front of the Winchester Square fire station around 1940. (Courtesy Connecticut Fire Museum archives.)

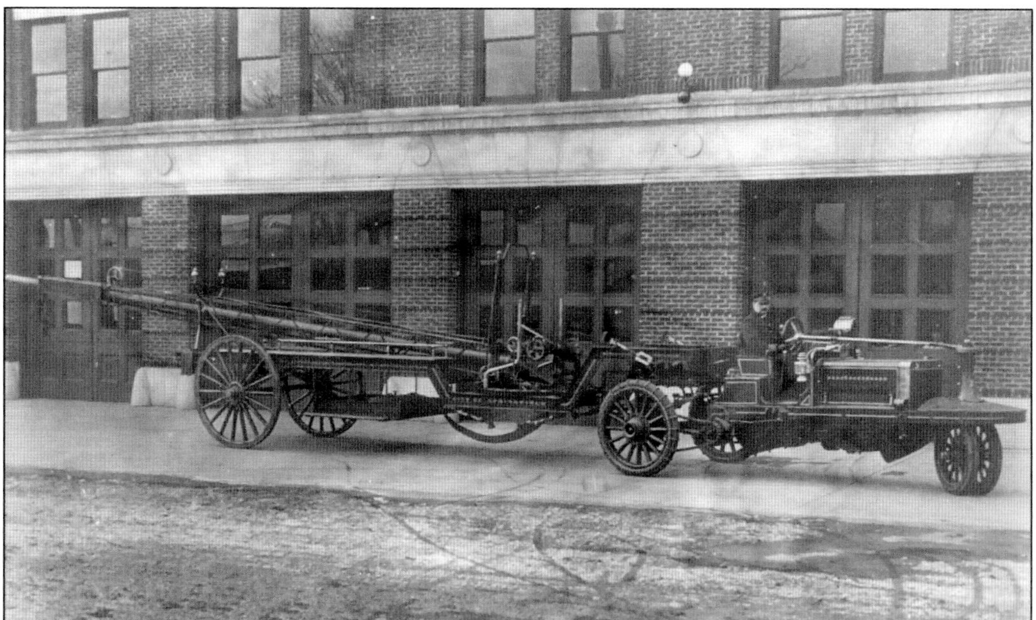

This is a view of the 1894 Champion water tower with its 1912 Knox-Martin tractor in front of the Court Street fire headquarters. The tower was delivered to the Pynchon Street fire station on February 5, 1894, and cost the city $4,500. In 1911, it was transferred to the new fire headquarters on Court Street. The original wagon wheels were replaced with rubber tires in 1939. An Autocar tractor replaced the Knox-Martin tractor in 1925 and was still in use when the tower was retired in 1955. (Courtesy Noah Pooler family archives.)

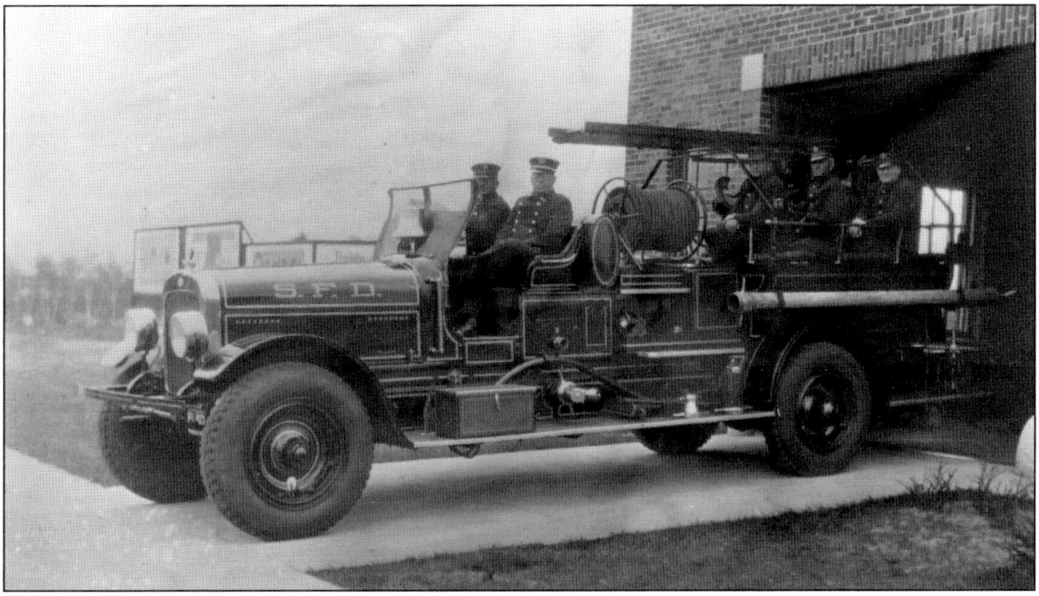

One of the 1930 Seagrave 1,000-gallon-per-minute pumpers is shown in front of the Page Boulevard firehouse in the early 1930s. Springfield purchased three pumpers from Seagrave in 1930 that were painted black rather than the more traditional red. This truck served until 1973, when it was sold at auction to the Connecticut Fire Museum in East Windsor, Connecticut. It was on display there until the 1980s, when it was sold to a collector in Great Britain where the truck remains today. (Courtesy Connecticut Fire Museum archives.)

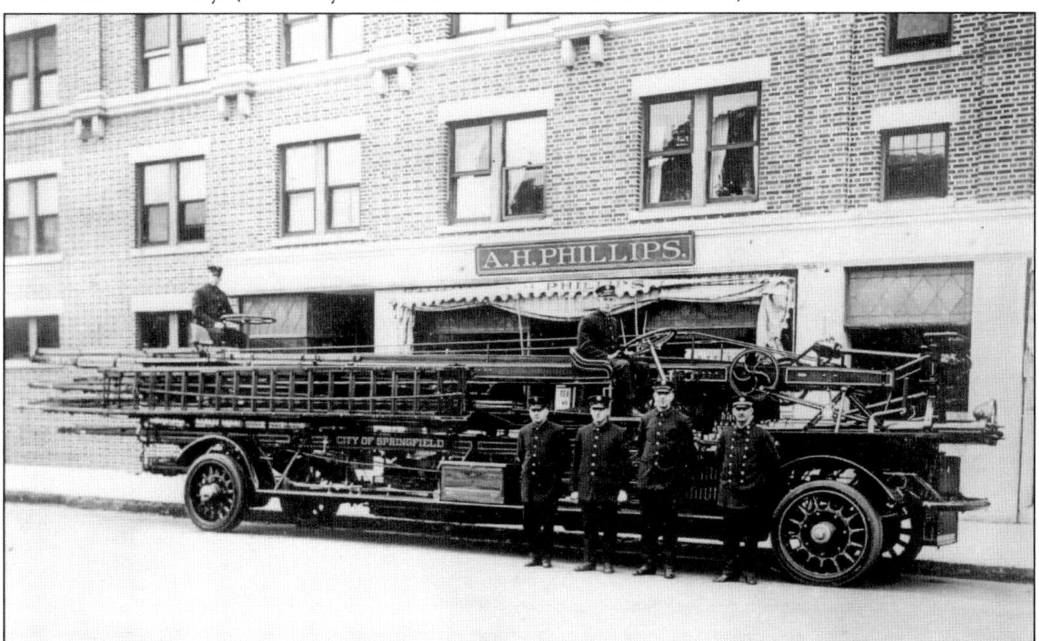

This 1910 Seagrave is a good example of the spring-raised aerial ladder the company developed in 1902 that took the muscle work out of raising the main ladder. A powerful spring lifted the bed ladder to a near vertical position, and the fly ladder was then extended by a large hand wheel. (Courtesy Connecticut Fire Museum archives.)

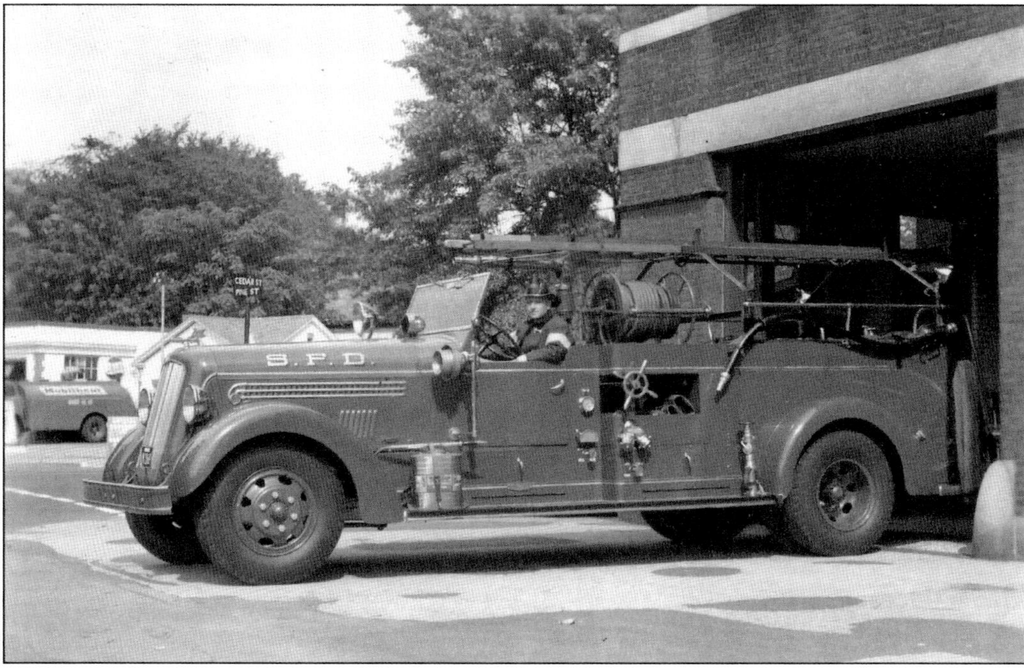

The 1939 Seagrave hose wagon of Hose Company No. 13 is shown in this 1950s view at the Pine Street firehouse. Springfield was one of the few fire departments in the country that actually used the number 13 for a company. (Courtesy Raymond Pond.)

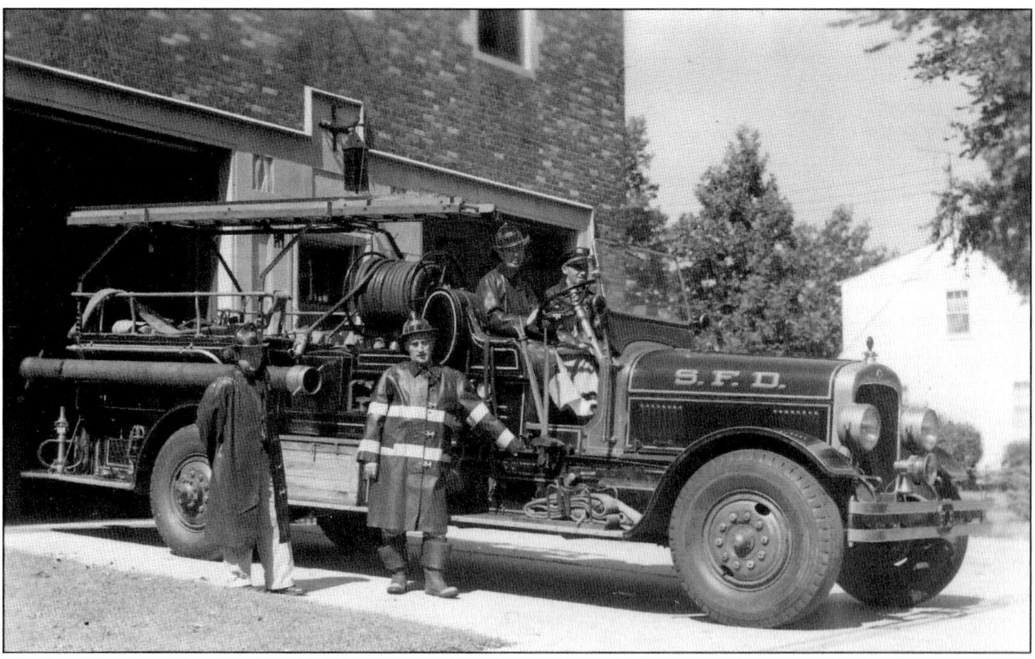

The 1930 Seagrave of Engine Company No. 16 is shown in front of the Massreco Street fire station. Because of the design of the squad body, Springfield's hose wagons and pumpers all had their roof ladders in an overhead ladder rack that gave the trucks a European look. (Courtesy Raymond Pond.)

Four

The Stations

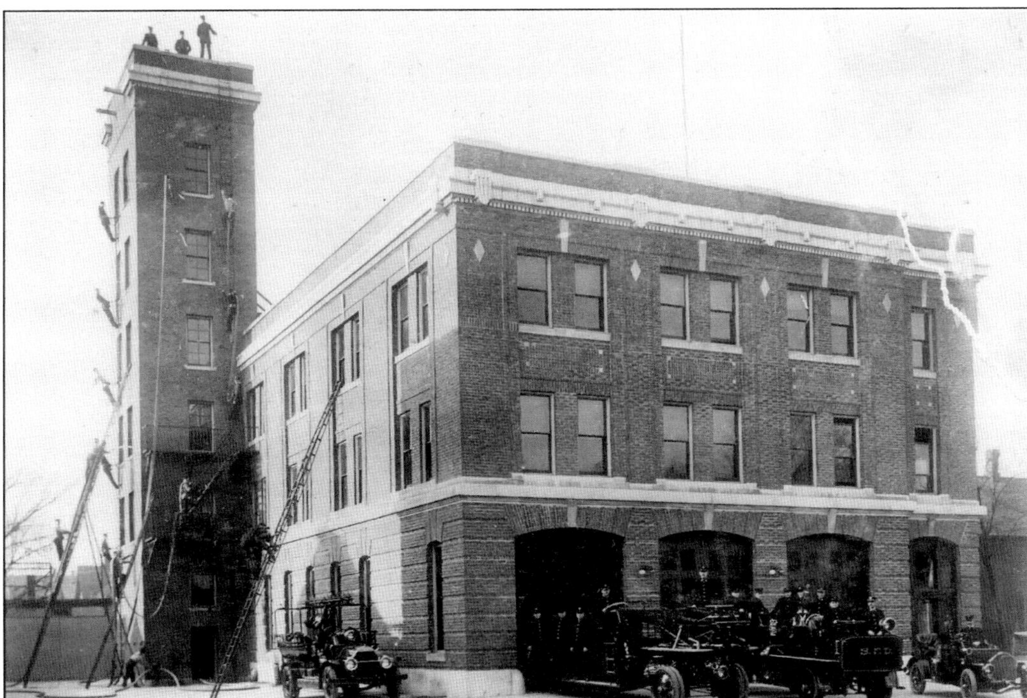

The North Street fire station, with its drill tower, opened in 1912. North Street was later renamed Dwight Street. Notice the firemen using their scaling ladders to climb the tower. For safety reasons, scaling ladders are no longer used by the fire service. (Courtesy Connecticut Fire Museum archives.)

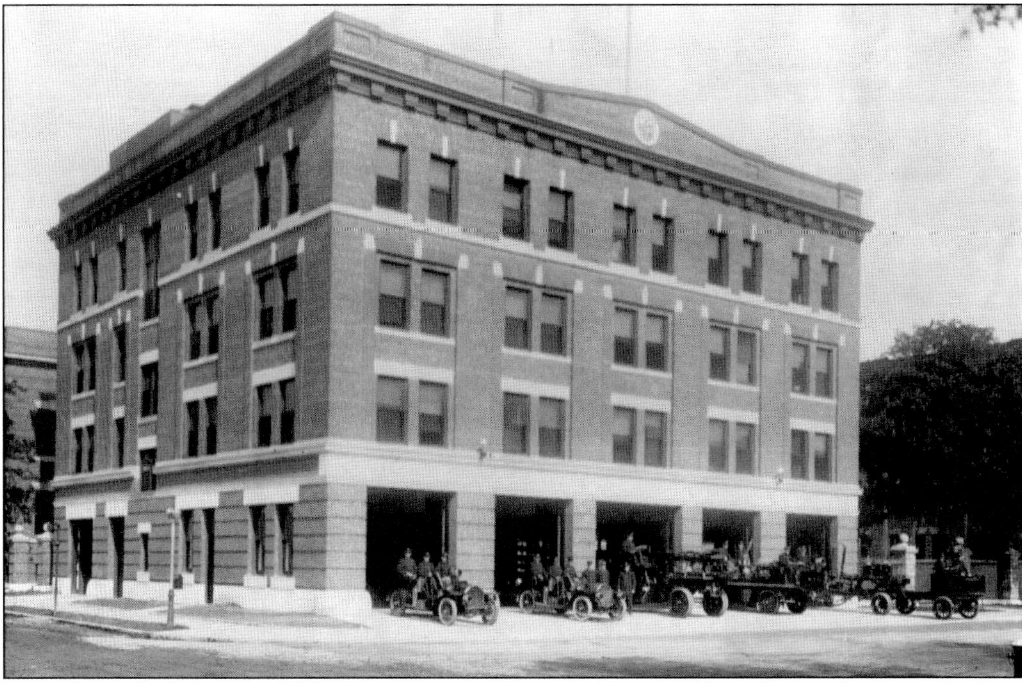

The fire headquarters at 86 Court Street is shown in this view from about 1920. (Courtesy Connecticut Fire Museum archives.)

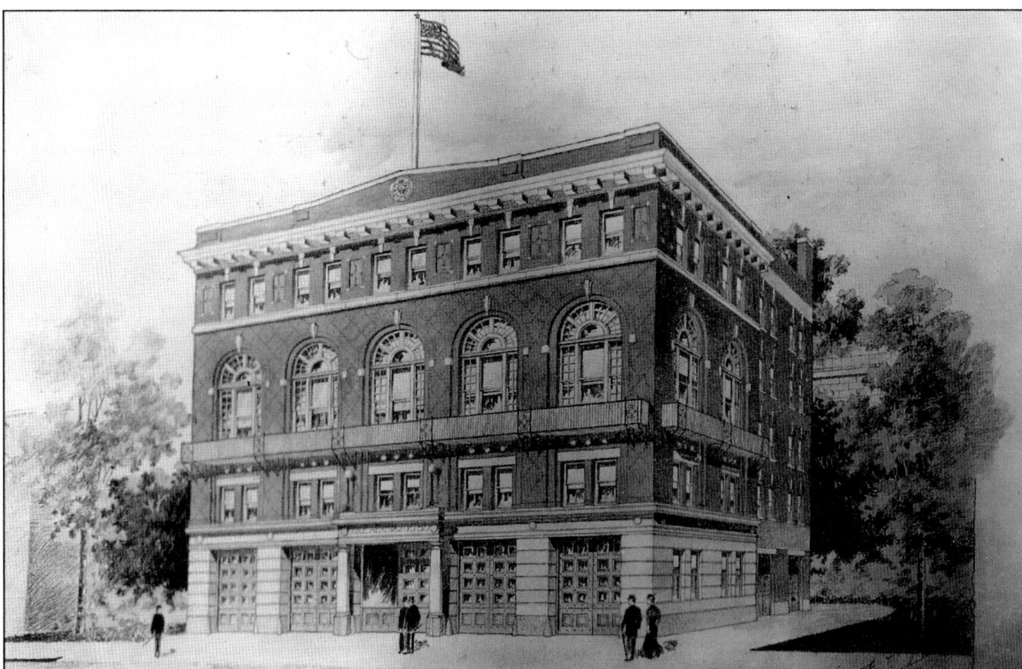

This is an architect's drawing of the new fire department headquarters at the intersection of Court Street and Columbus Avenue that was opened in 1911. (Courtesy Connecticut Fire Museum archives.)

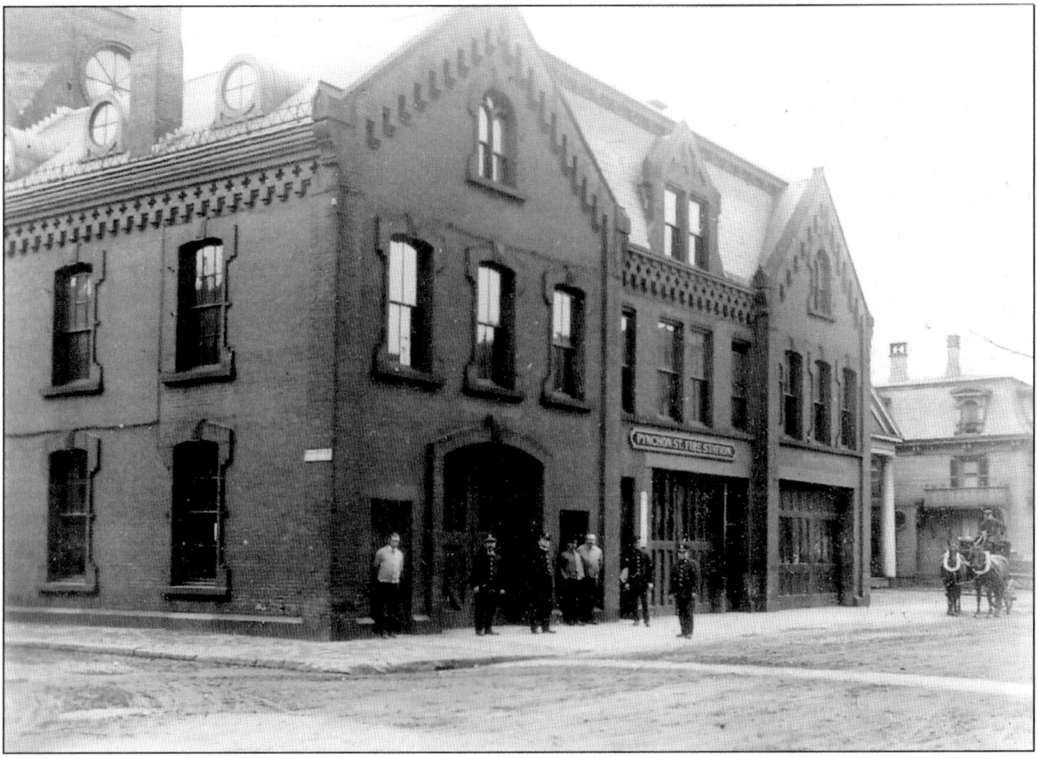

Pynchon Street fire station, built in 1864, served as the fire department headquarters and also housed Hose Company No. 1, Ladder Company No. 1 and Chemical Company B. (Courtesy Noah Pooler Family archives.)

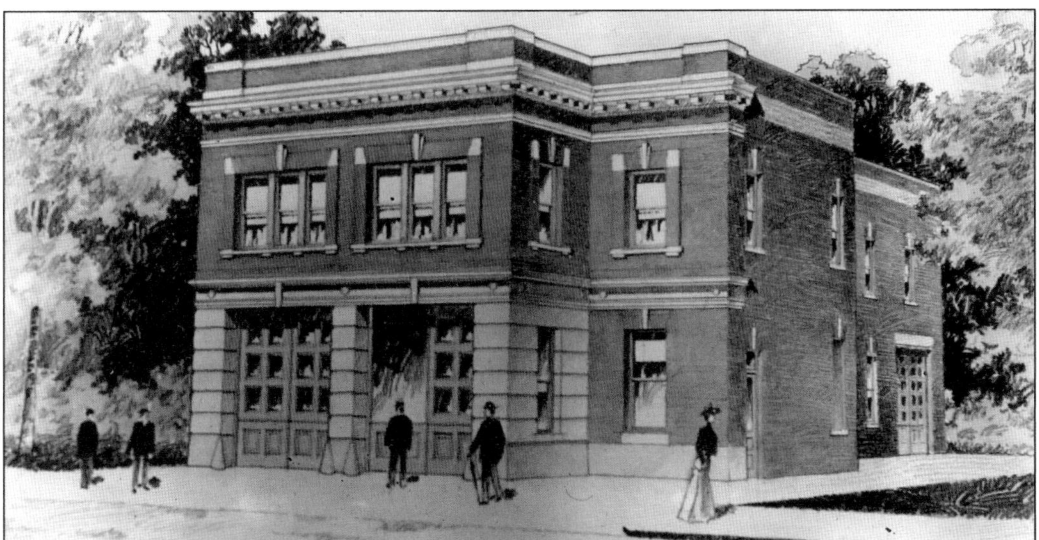

This is an architect's drawing of a new station that was built at 56 Margaret Street. B. Hammet Seabury designed many of the firehouses built in Springfield during the 1920s. This was the first station built in Springfield to house motor fire apparatus. This building now is used for the Lady of Mount Carmel preschool. (Courtesy Noah Pooler Family archives.)

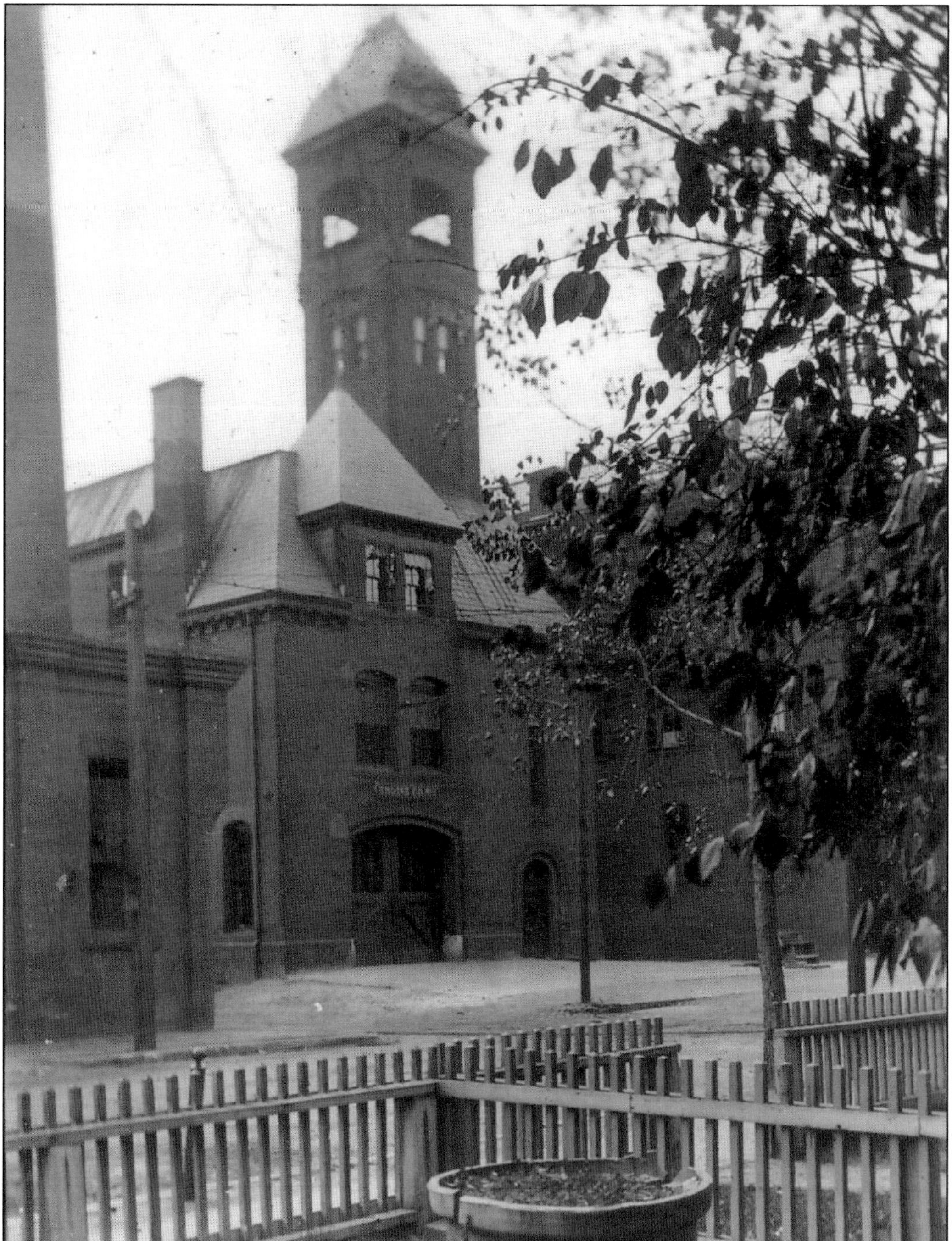

Bond Street fire station housed Hose Company No. 2 in 1910. The athletic field for Commerce High School occupies this site today. (Courtesy Noah Pooler family archives.)

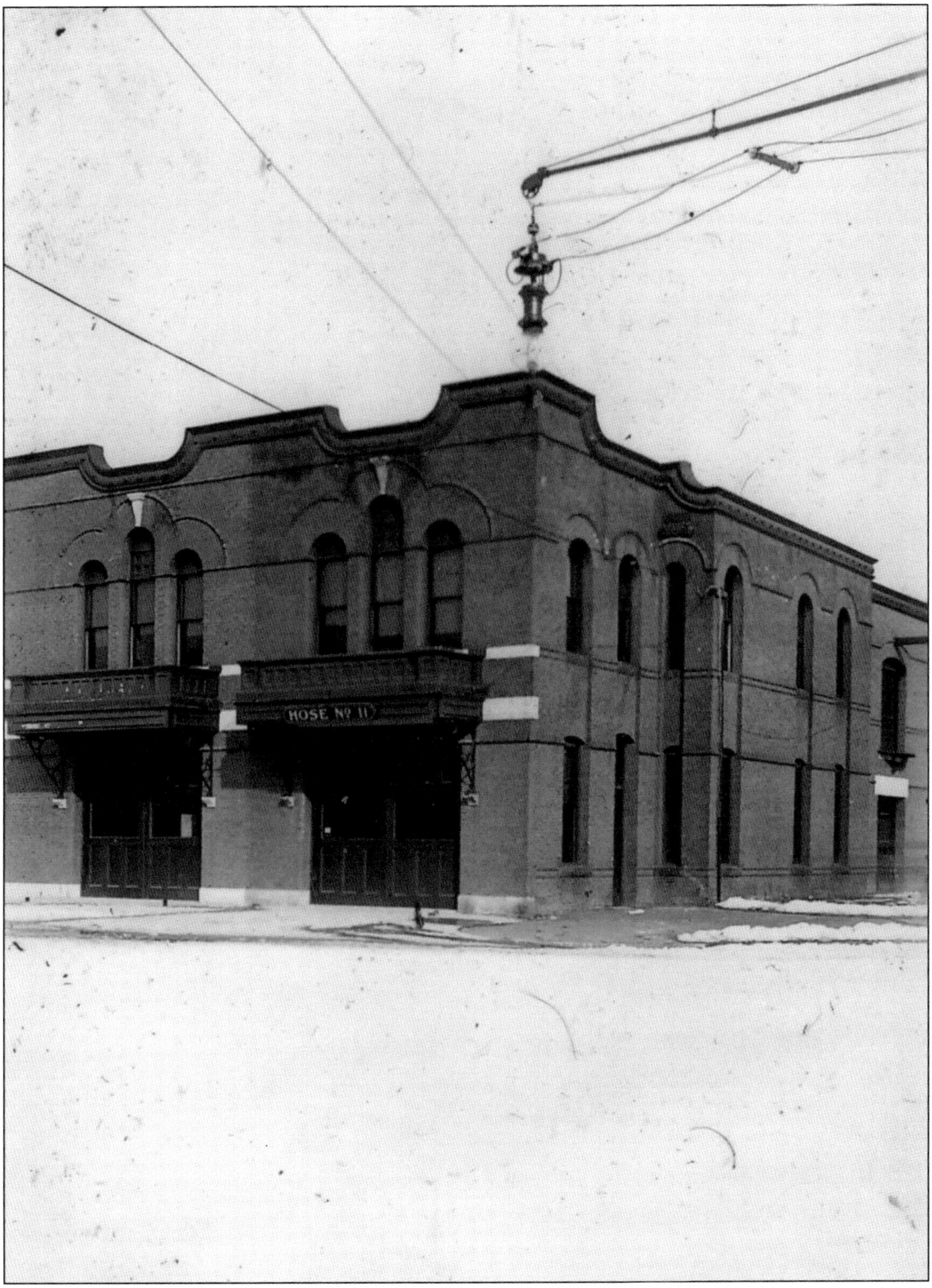

Oakland Street fire station was originally built to house Chemical Company E, and later Hose Company No. 11 operated out of this facility. For the last 90 years, it has been the quarters of Engine Company No. 3. (Courtesy Noah Pooler family archives.)

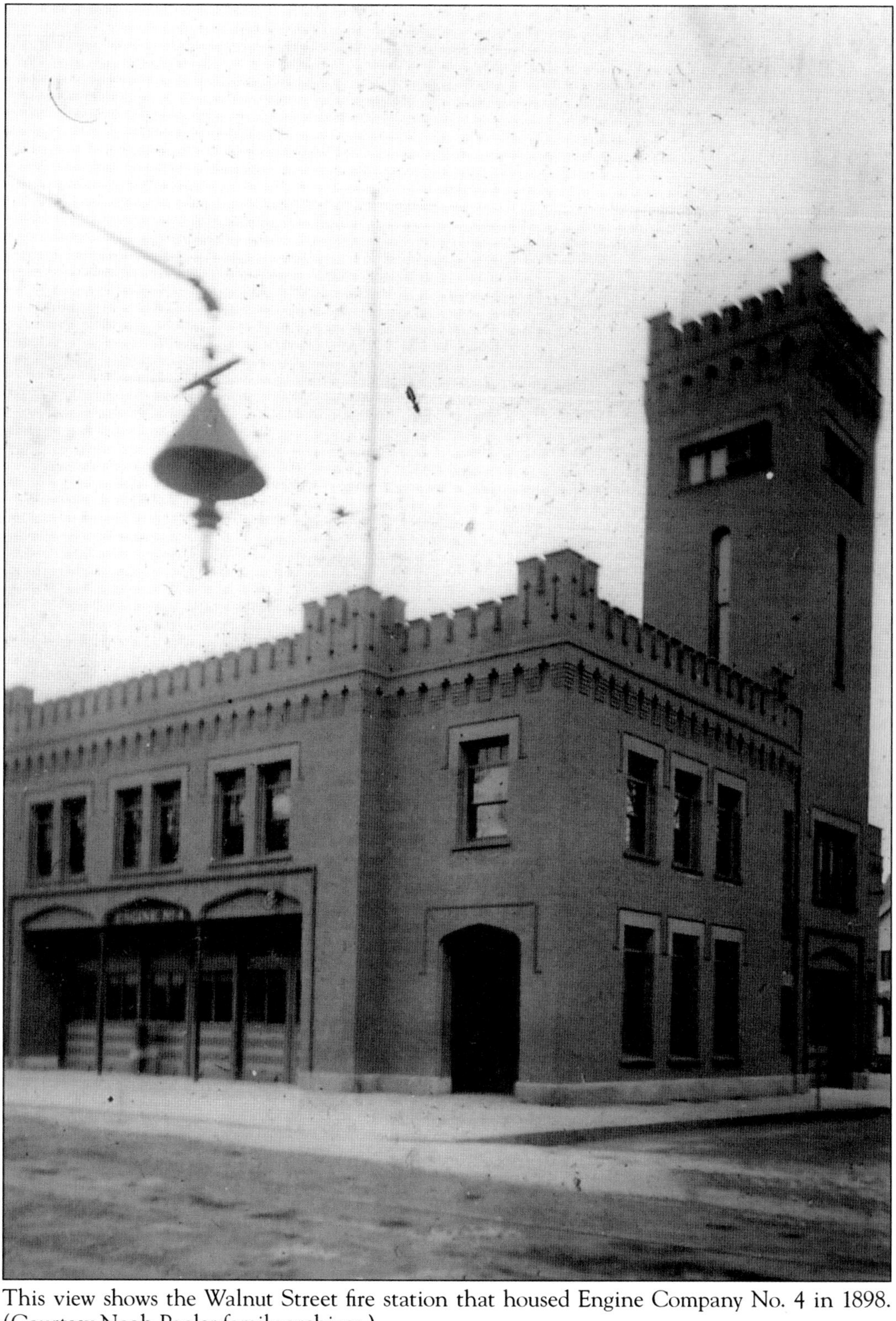

This view shows the Walnut Street fire station that housed Engine Company No. 4 in 1898. (Courtesy Noah Pooler family archives.)

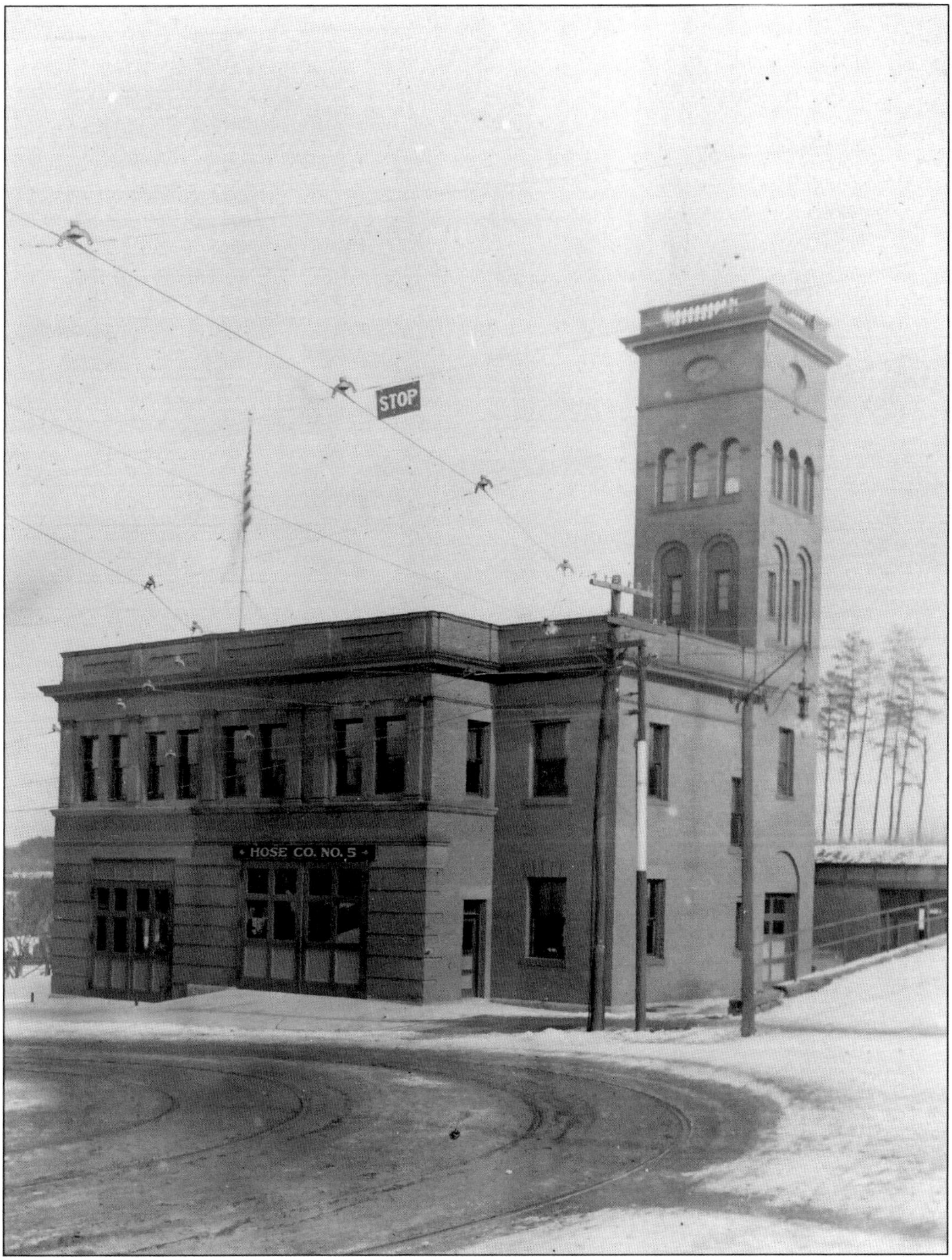

The Indian Orchard fire station on Oak Street is shown in 1909. The tower in the rear of the building was used for drying wet fire hose. Canvas fire hose had to be air dried after getting wet or mildew would begin to grow on the canvas. In 1914, a streetcar ran into the building after the motorman lost control coming around the corner. The fire station building still stands and is currently used by the board of education as a storage building. (Courtesy Noah Pooler family archives.)

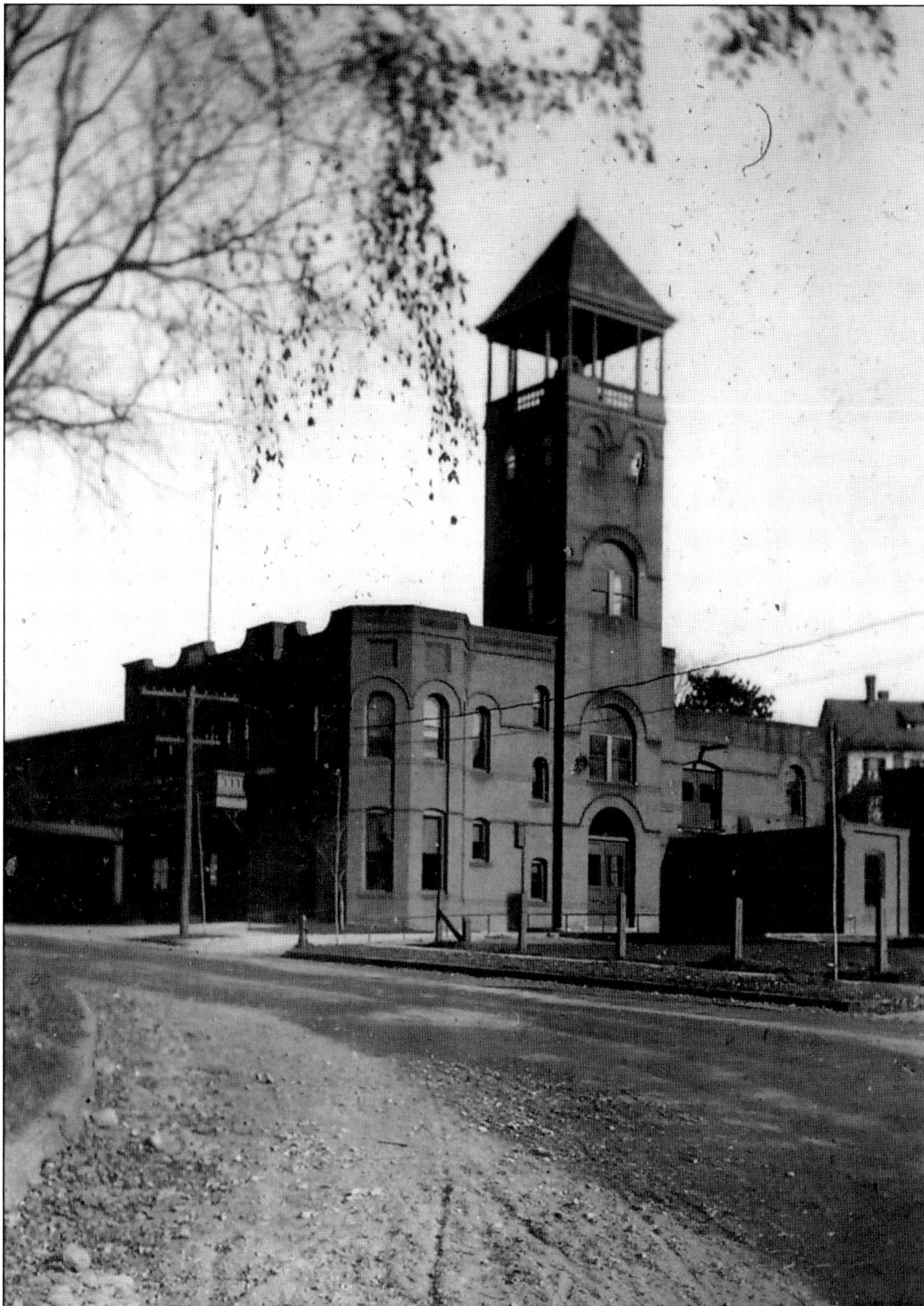

This view shows the Mill Street fire station as it looked in the late 1890s. Note that Mill Street was still a dirt road. (Courtesy Noah Pooler family archives.)

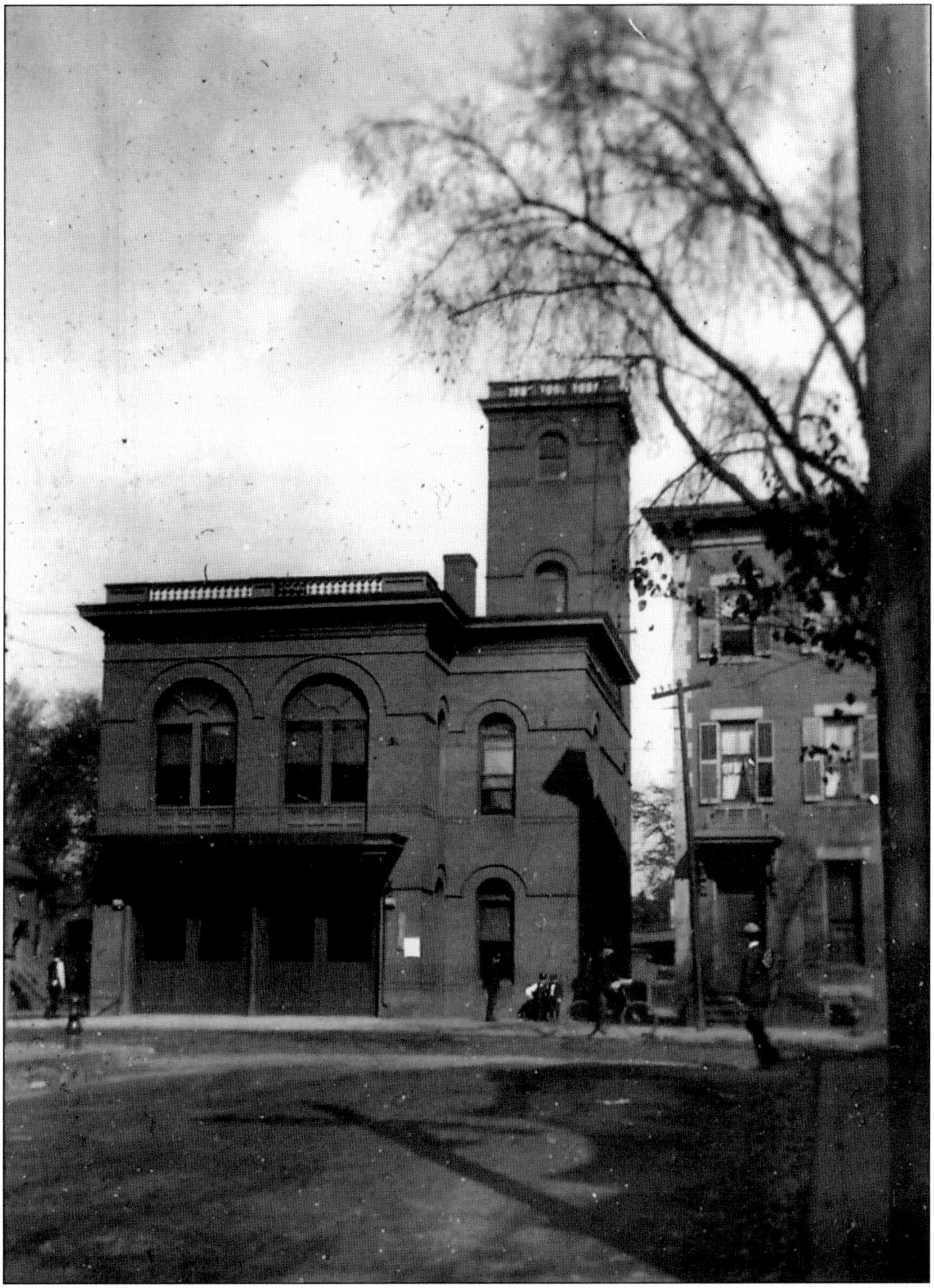

The Chestnut Street fire station is shown in this 1900 view. Note the hose-drying tower in the rear of the building. This was standard on most fire stations until the 1960s, when hose drying machines came into use. When this building was torn down, Taylor Street was extended through the site. (Courtesy Noah Pooler family archives.)

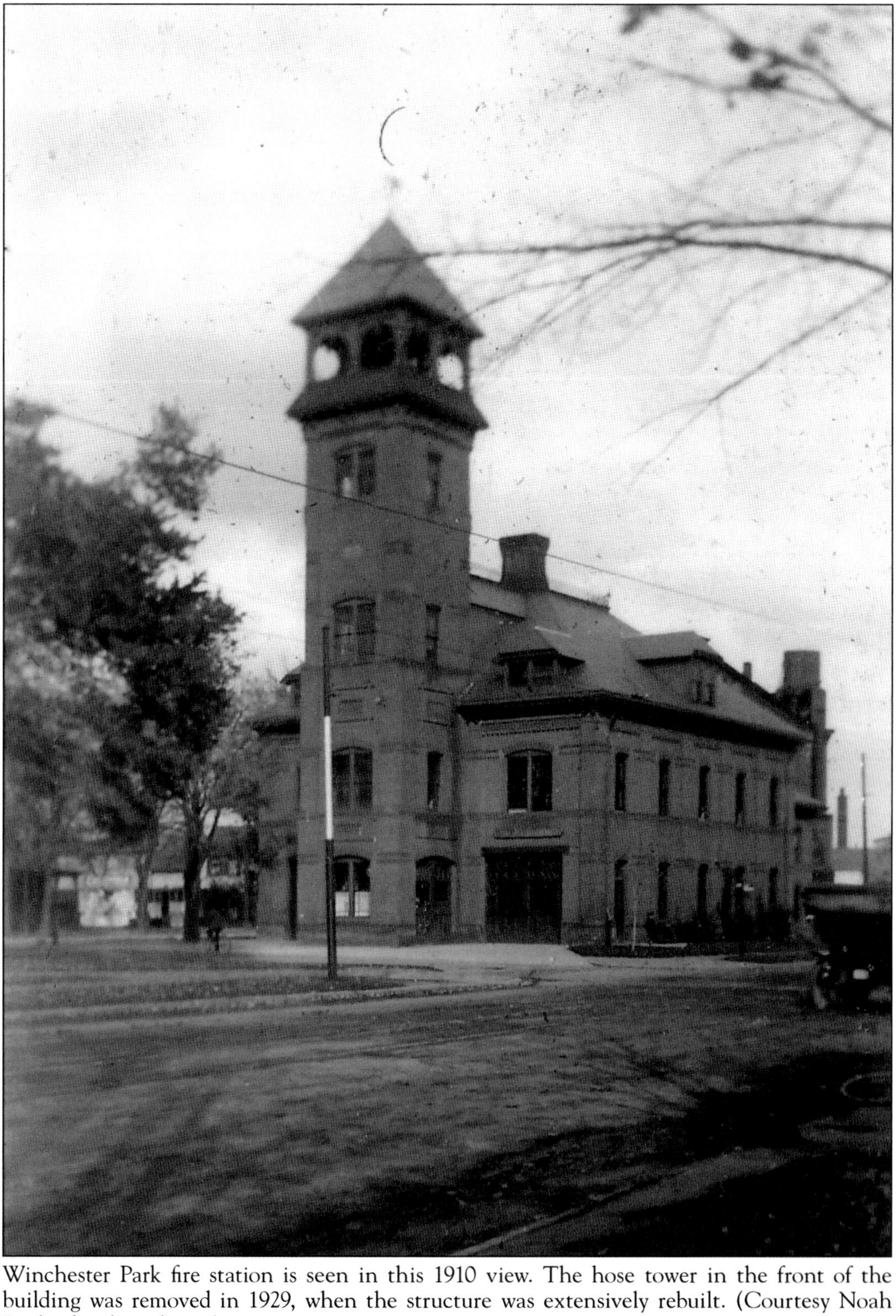

Winchester Park fire station is seen in this 1910 view. The hose tower in the front of the building was removed in 1929, when the structure was extensively rebuilt. (Courtesy Noah Pooler family archives.)

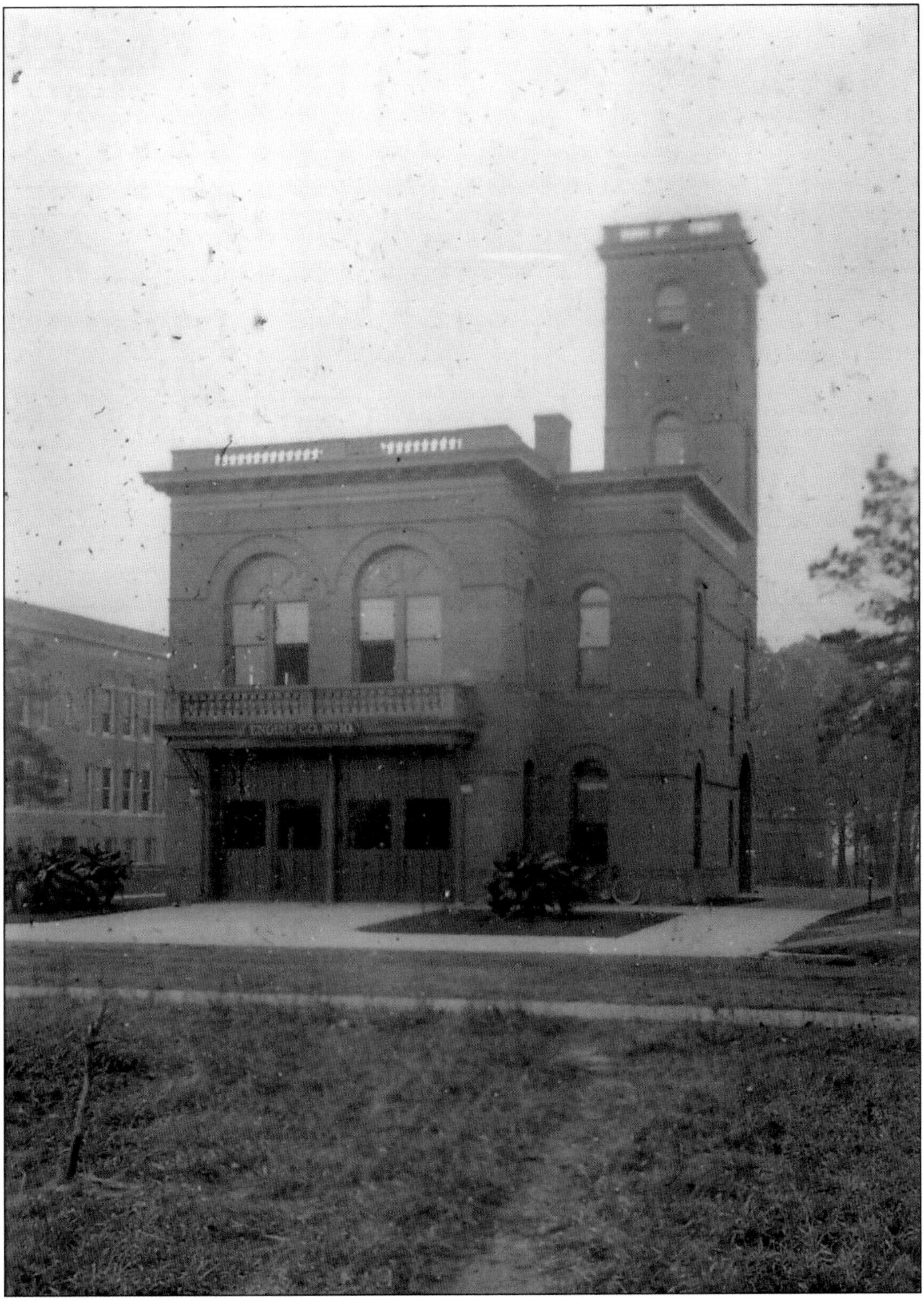

This building at 438 Armory Street housed Engine Company No. 10. It was nearly identical to the Chestnut Street station and was in service from 1902 until 2002. (Courtesy Noah Pooler family archives.)

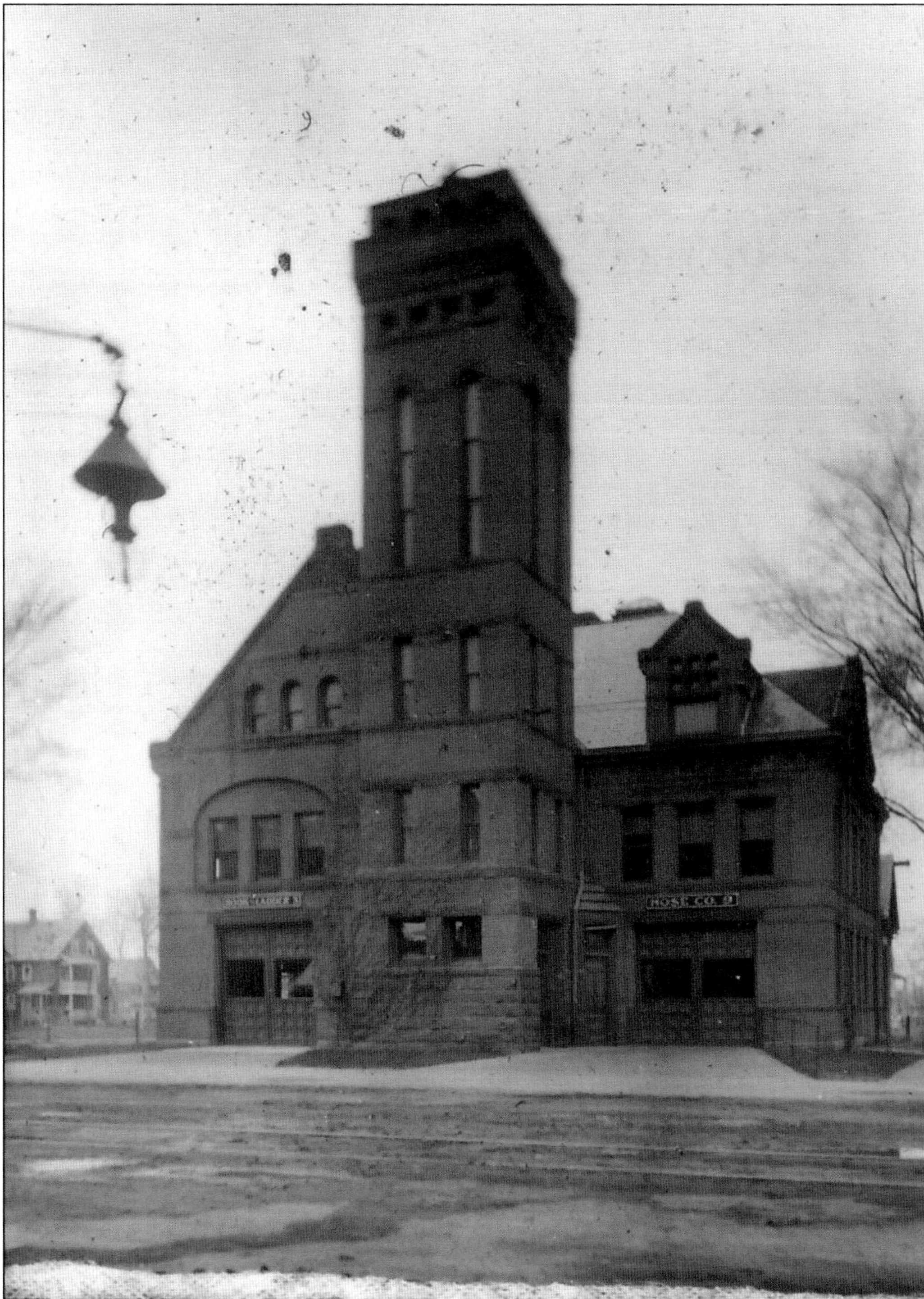

The North Main Street fire station had a hose tower in the front of the building like Winchester Park. This station housed Hose Company No. 9 and Ladder Company No. 3 in 1911. (Courtesy Noah Pooler family archives.)

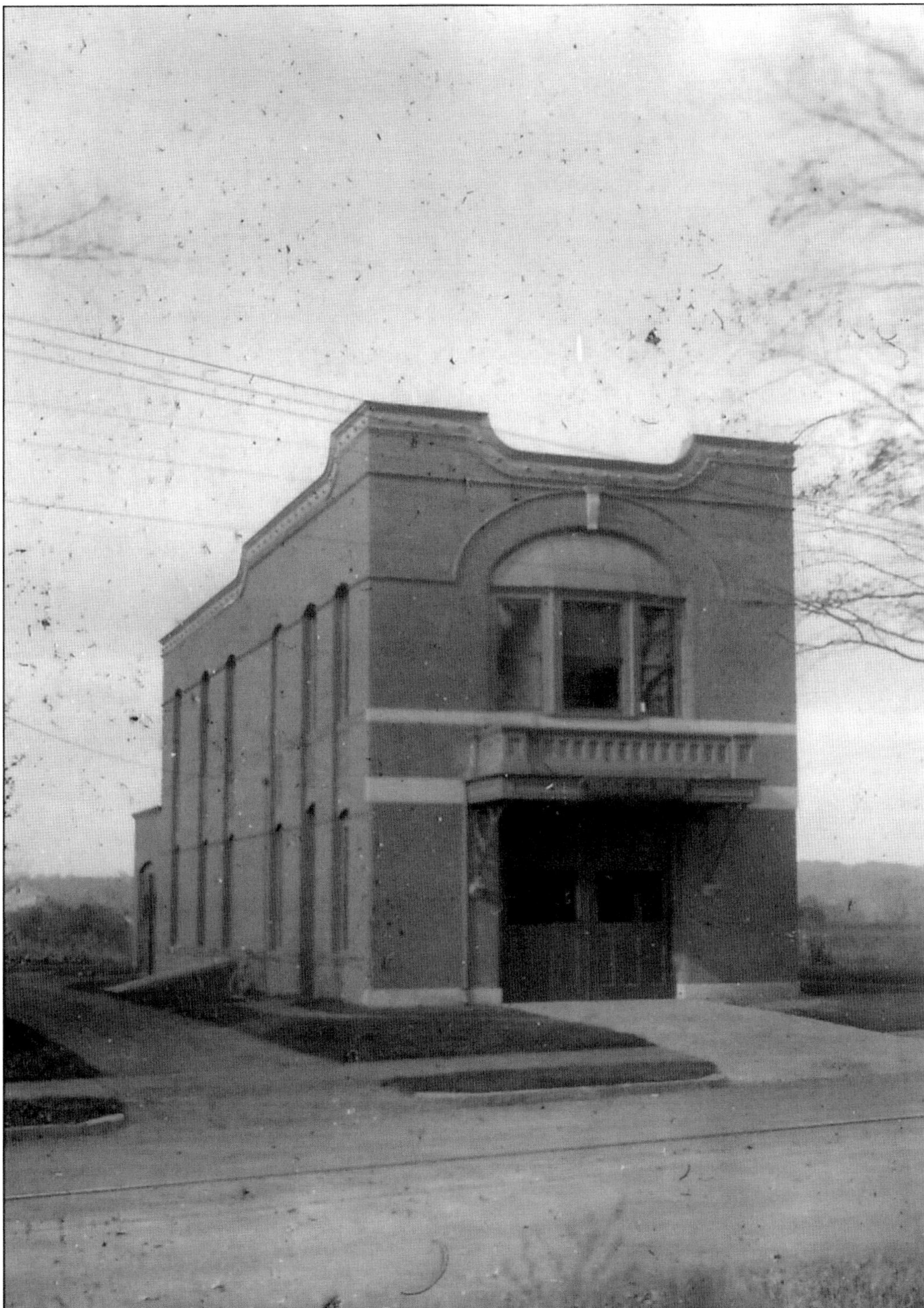

This view shows the Plainfield Street station that looks like it was built in a rural part of the city. This housed Hose Company No. 13 in 1910. (Courtesy Noah Pooler family archives.)

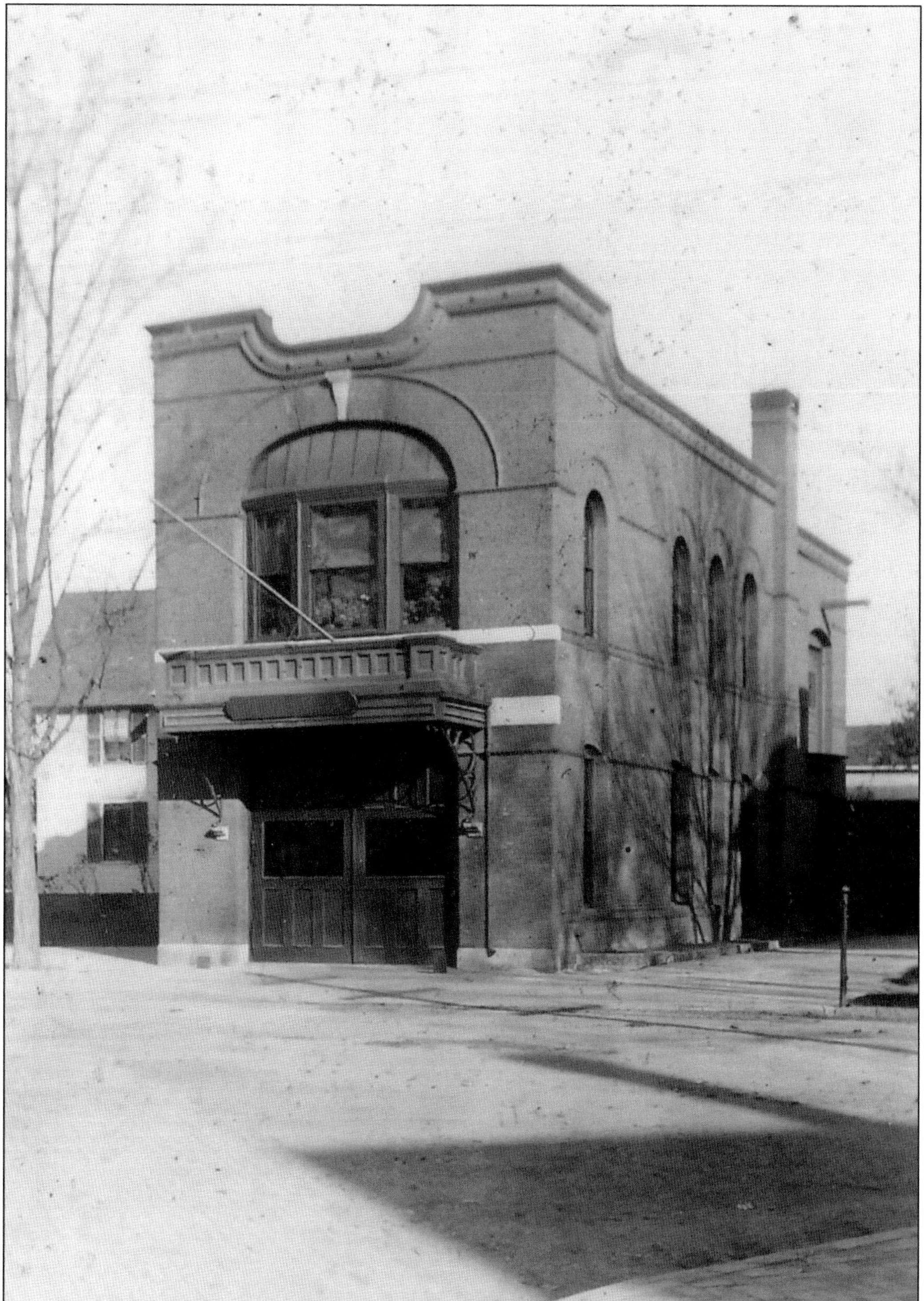

The station at 147 Pine Street was originally built as a one-bay station for Chemical Company A. (Courtesy Noah Pooler family archives.)

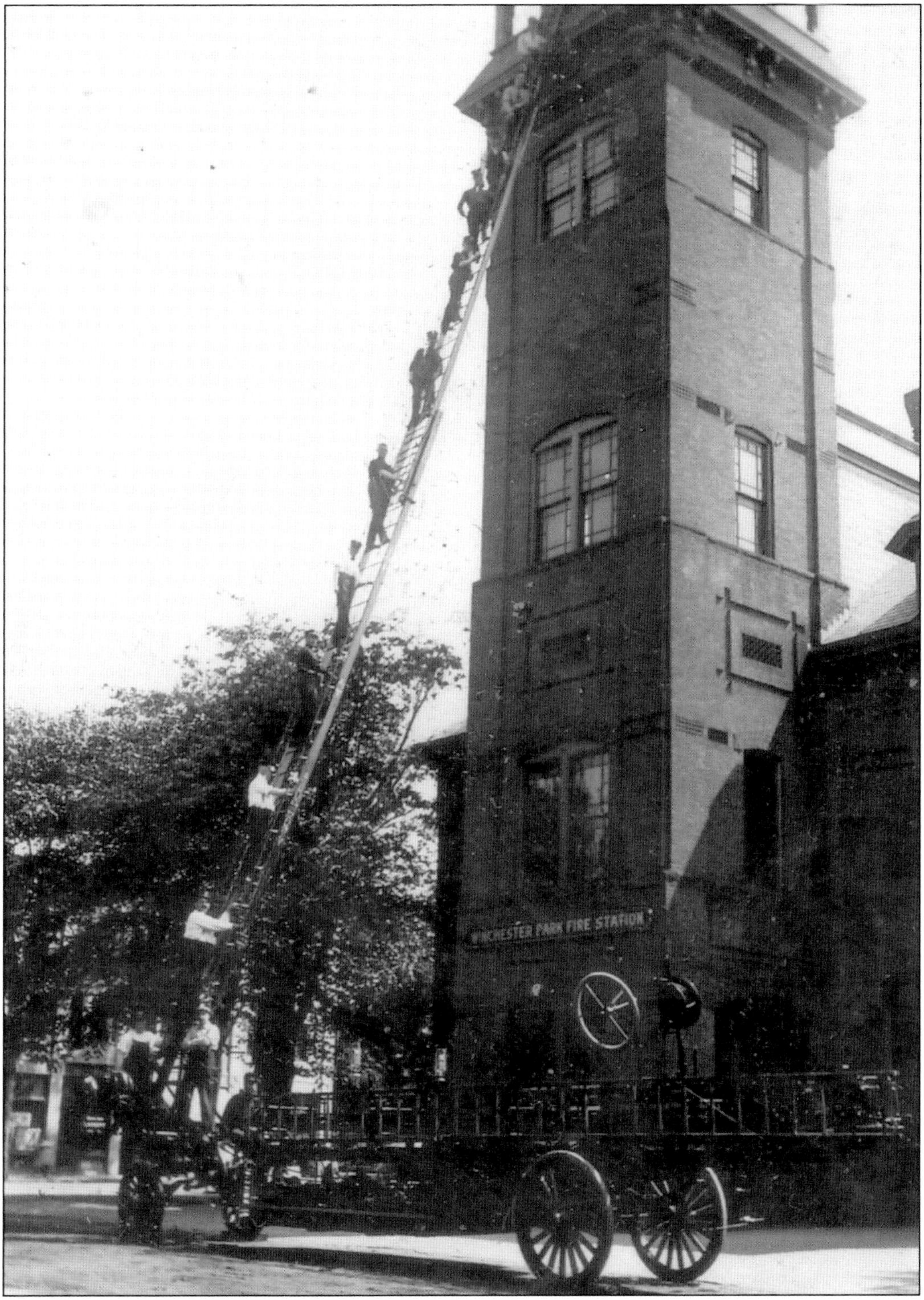

Ladder Company No. 8 operated out of the Winchester Park fire station. This 1905 view shows the company drilling. Note that the tiller-man steering wheel had to be removed before the ladder could be raised. (Courtesy Noah Pooler family archives.)

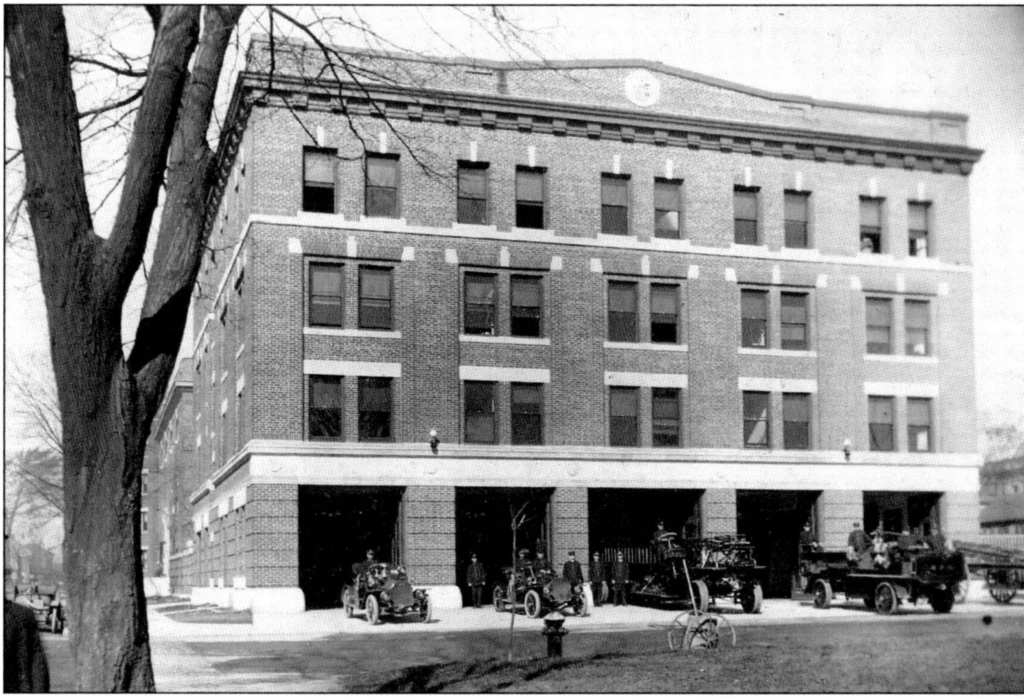

The new fire headquarters at the intersection of Court Street and Columbus Avenue is shown in this 1914 view. This building was demolished when Interstate 91 was built through Springfield in the early 1970s. (Courtesy Noah Pooler family archives.)

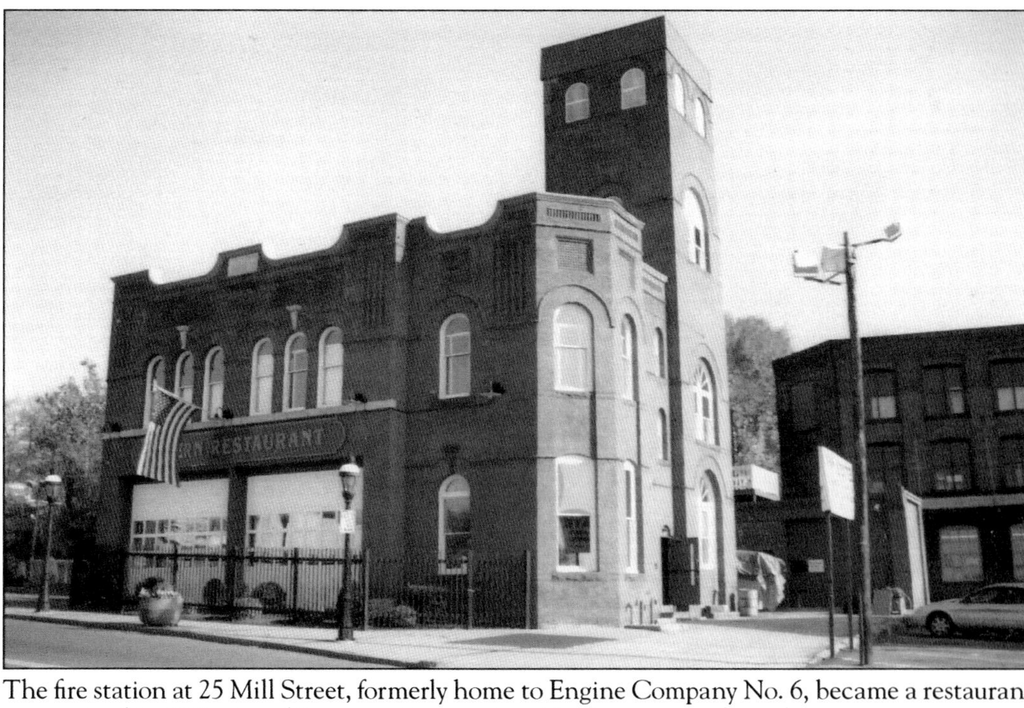

The fire station at 25 Mill Street, formerly home to Engine Company No. 6, became a restaurant as seen in this 2006 view. (Courtesy Connecticut Fire Museum archives.)

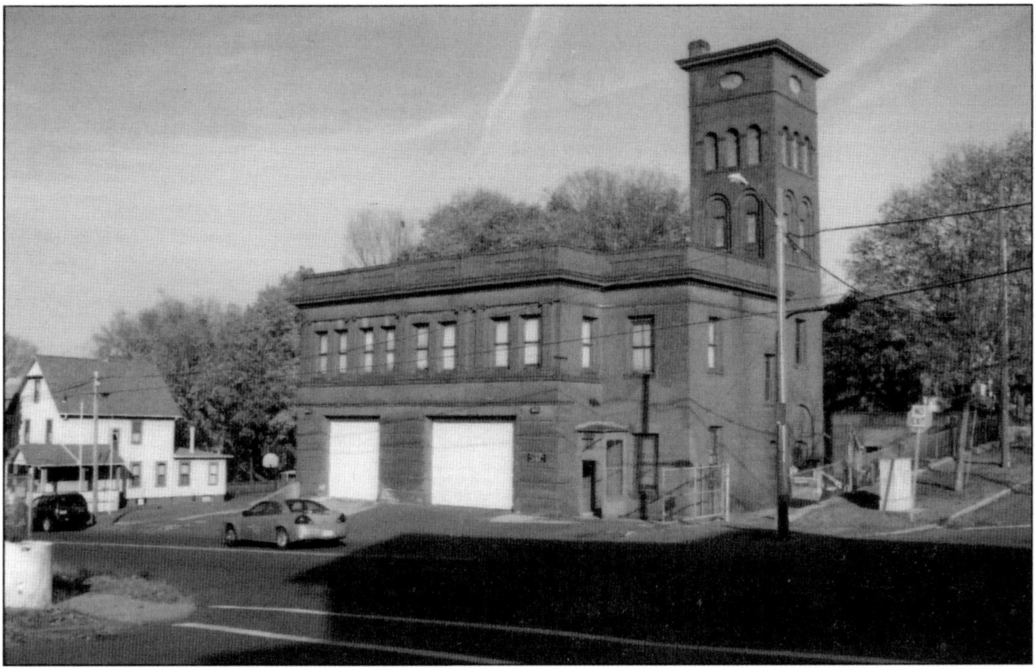

This is a 2006 view of the Indian Orchard fire station located at 97 Oak Street. It is now used as a warehouse by the Springfield Board of Education. Compare this to the picture on page 85 taken in 1909. (Courtesy Connecticut Fire Museum archives.)

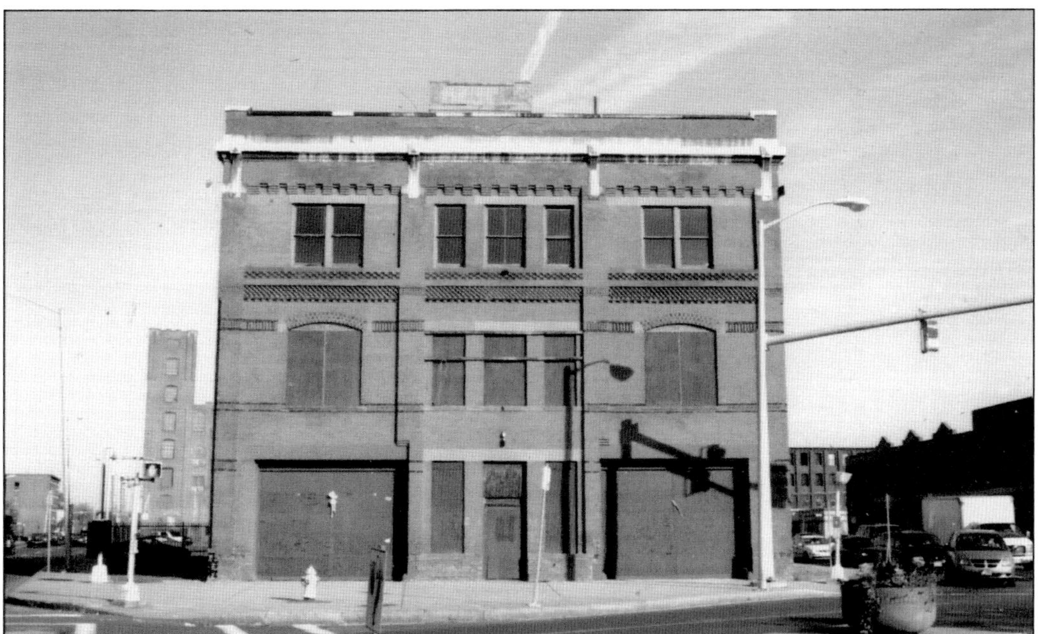

The fire station at Winchester Square is no longer used by the fire department. It housed Engine Company No. 8 and Ladder Company No. 8. (Courtesy Connecticut Fire Museum archives.)

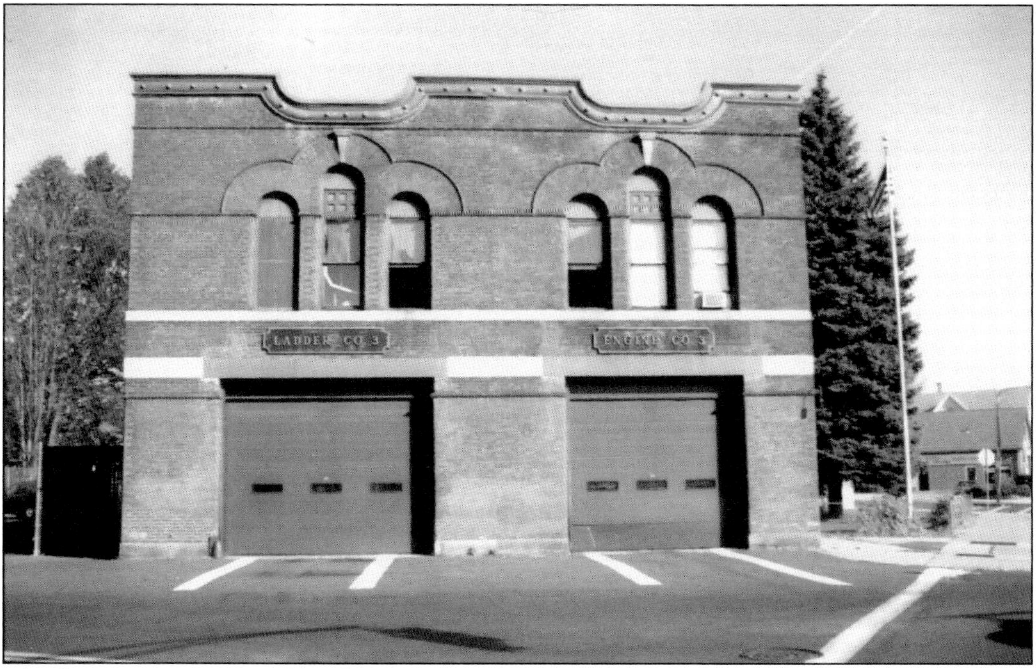

The fire station at 173 Oakland Street is still in service and is the home of Engine Company No. 3. Built in 1894, it has seen over 110 years of service to the city. (Courtesy Connecticut Fire Museum archives.)

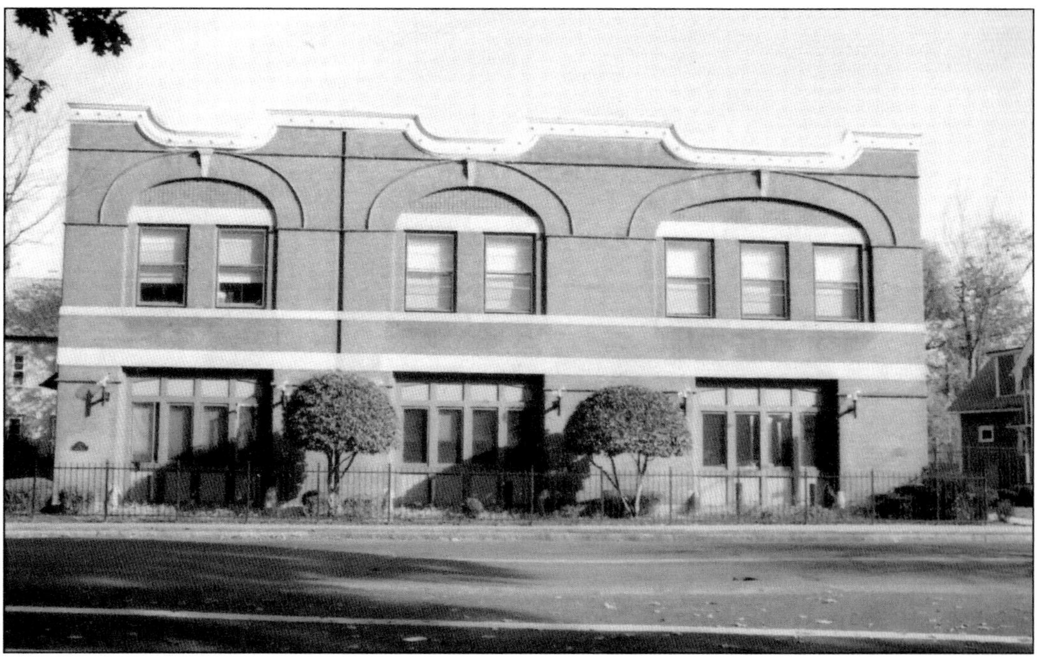

The Pine Street fire station is now a community center after serving as a fire station for over 75 years. (Courtesy Connecticut Fire Museum archives.)

Five

THE FIREMEN

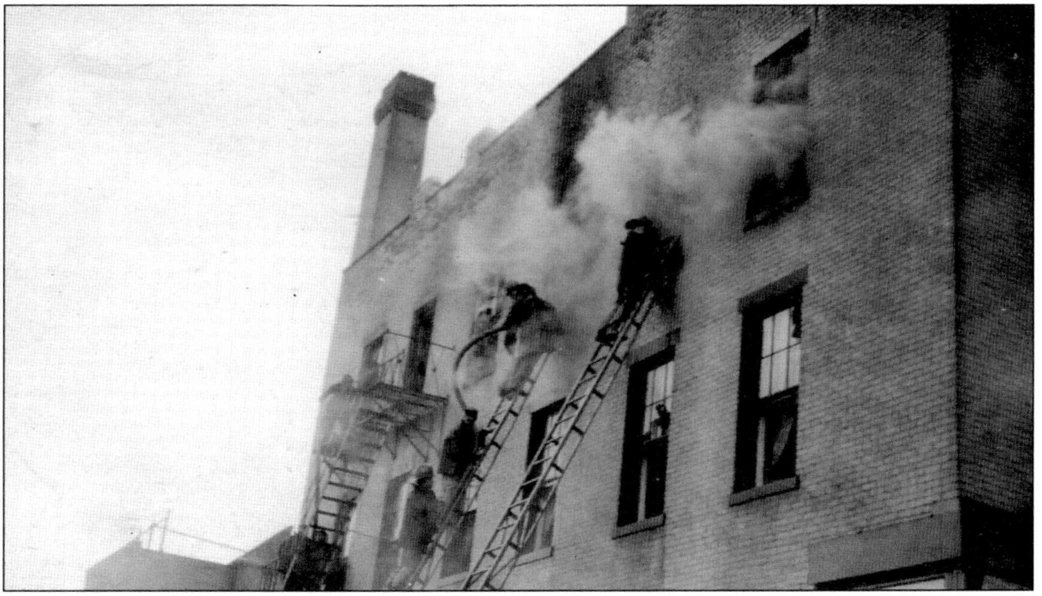

In 1915, Springfield's firemen had little protection from the hazards of firefighting like they do today. Rubber coats and boots were the normal protection. Only the leather fire helmets they are wearing are still in use today. In 1915, the department had 187 members including mechanics, linemen, dispatchers, the chief, and his deputies. (Courtesy Noah Pooler family archives.)

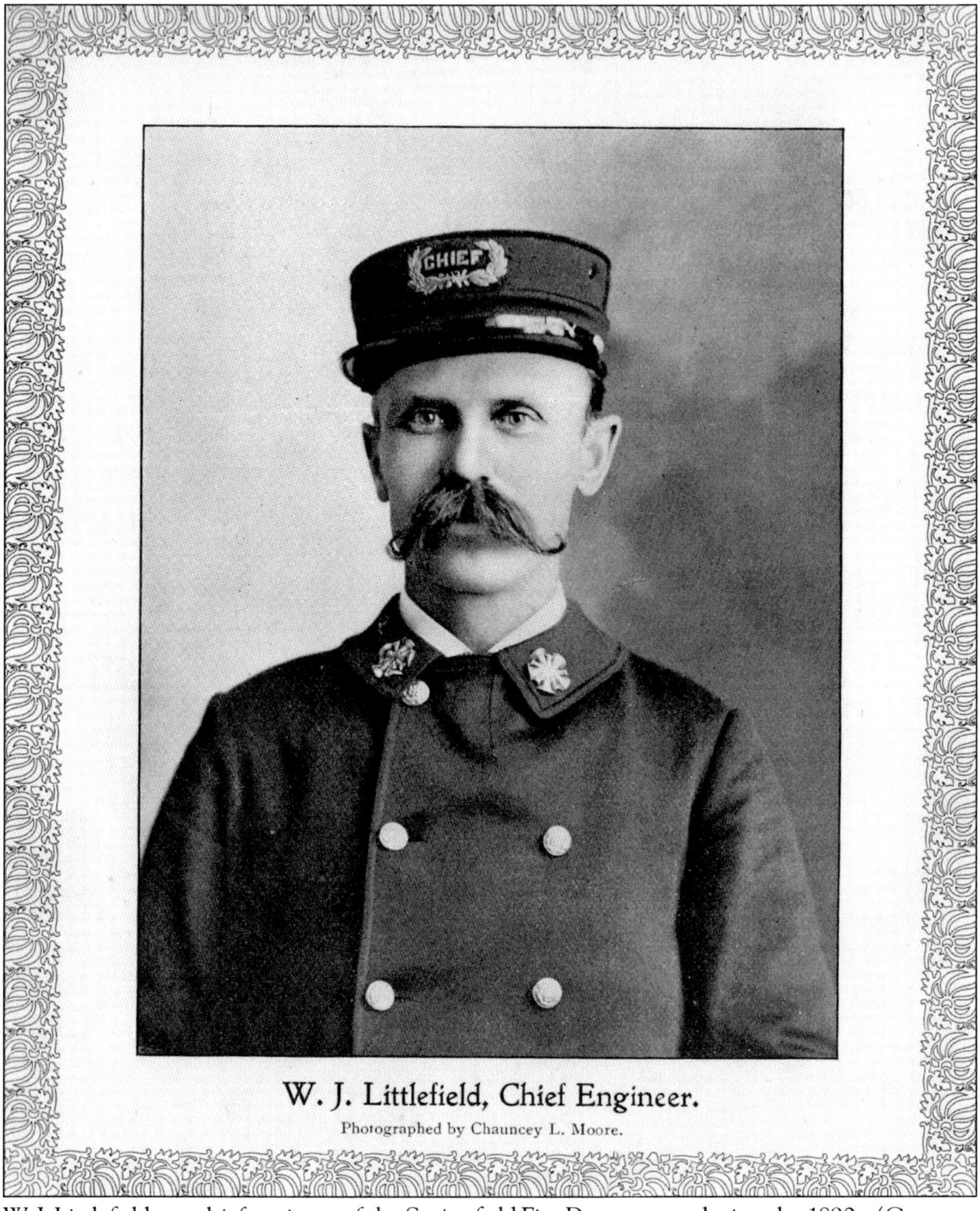

W. J. Littlefield was chief engineer of the Springfield Fire Department during the 1890s. (Courtesy Connecticut Fire Museum archives.)

Chief Littlefield lived in this house during the time he was the head of the Springfield Fire Department. (Courtesy Raymond Pond.)

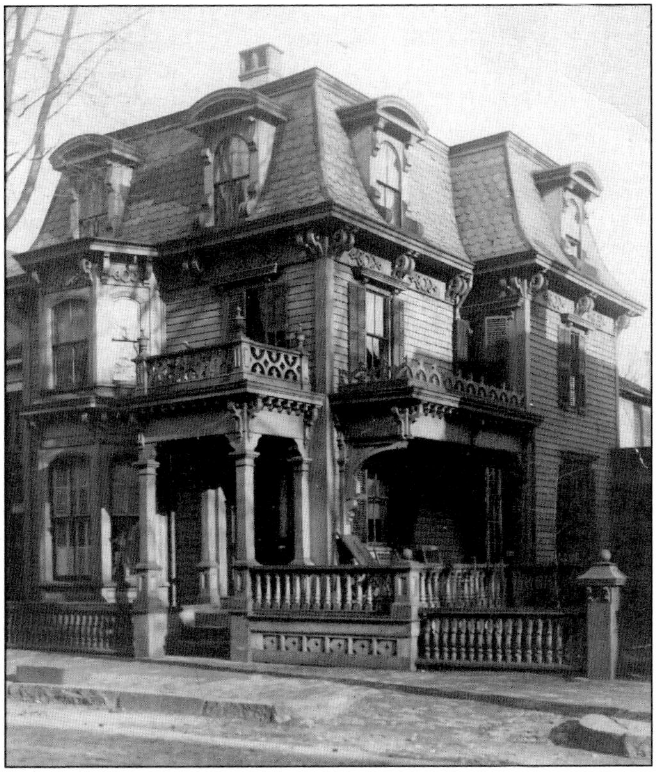

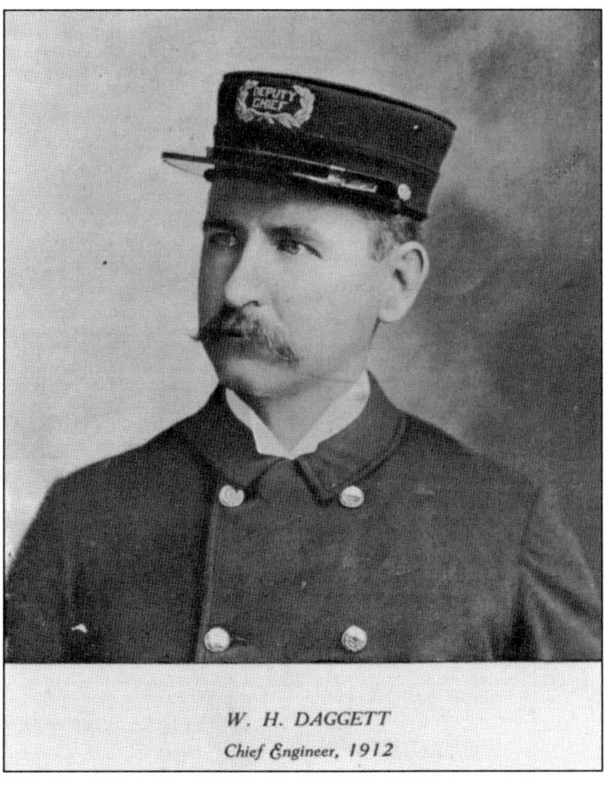

William H. Daggett was chief of the department in 1912. (Courtesy Connecticut Fire Museum archives.)

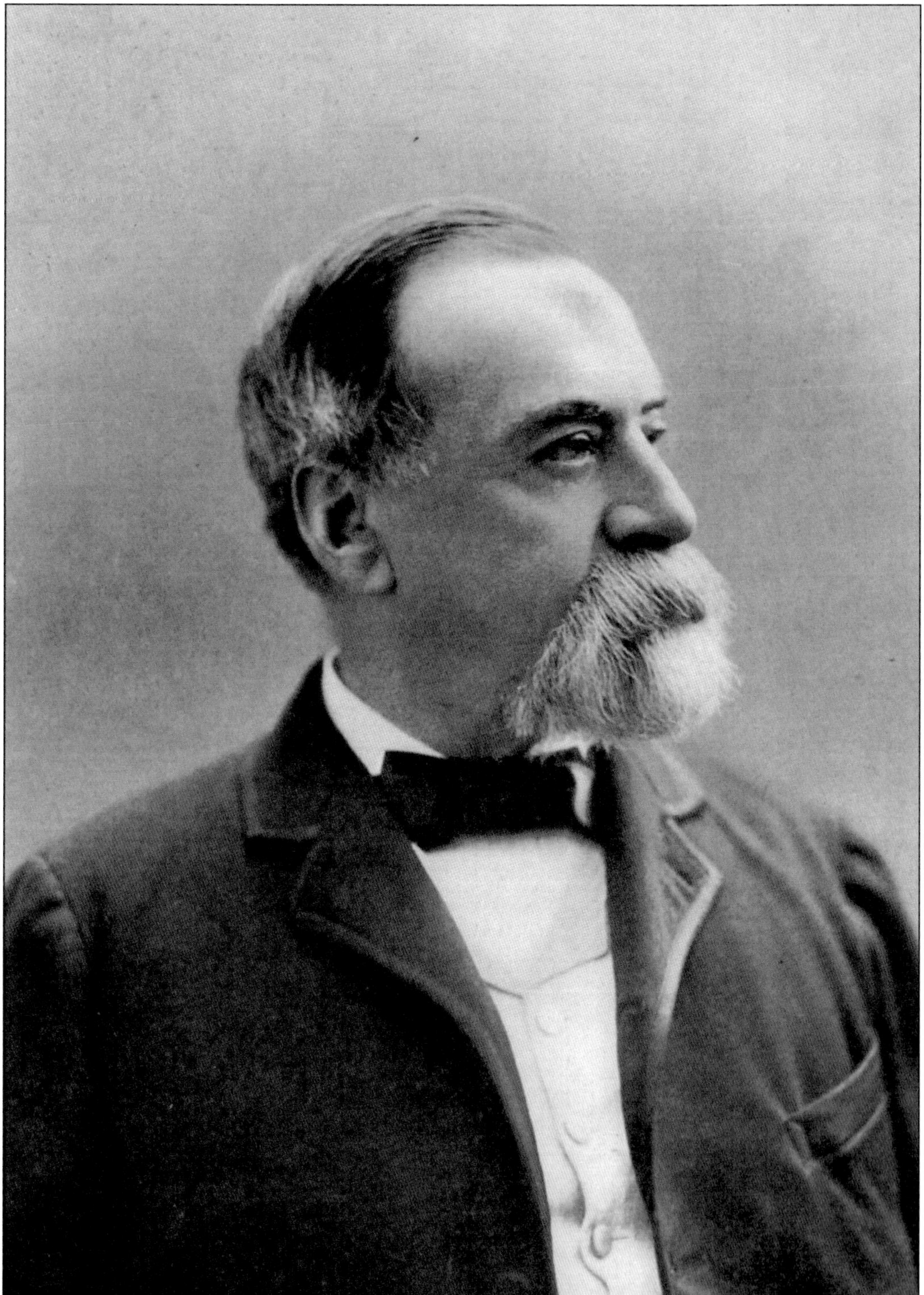

Abner P. Leshure was the chief engineer from 1870 to 1894. (Courtesy Connecticut Fire Museum archives.)

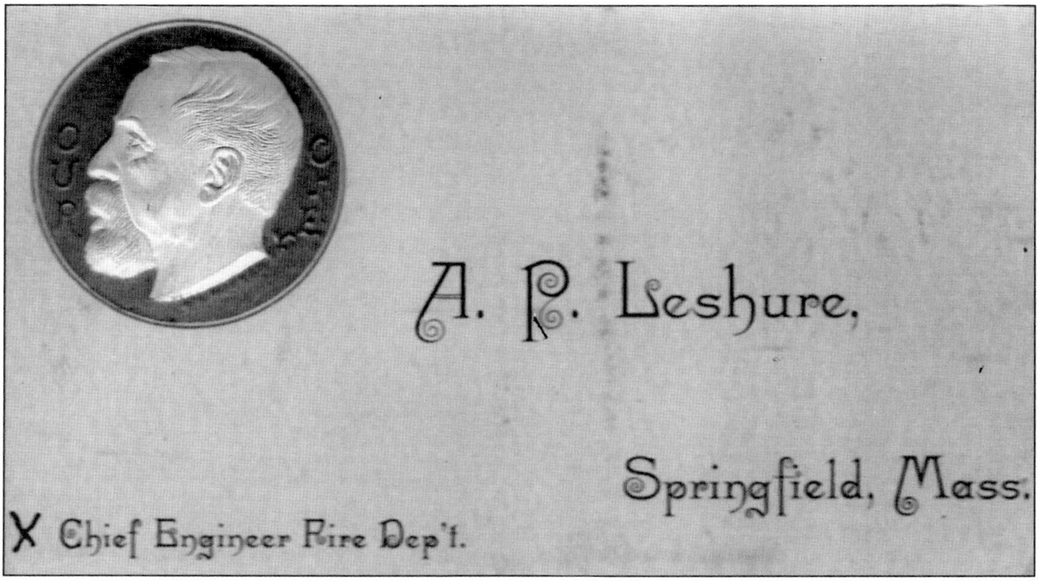

Pictured here is Chief Leshure's business card. (Courtesy Geoffery Neilsen.)

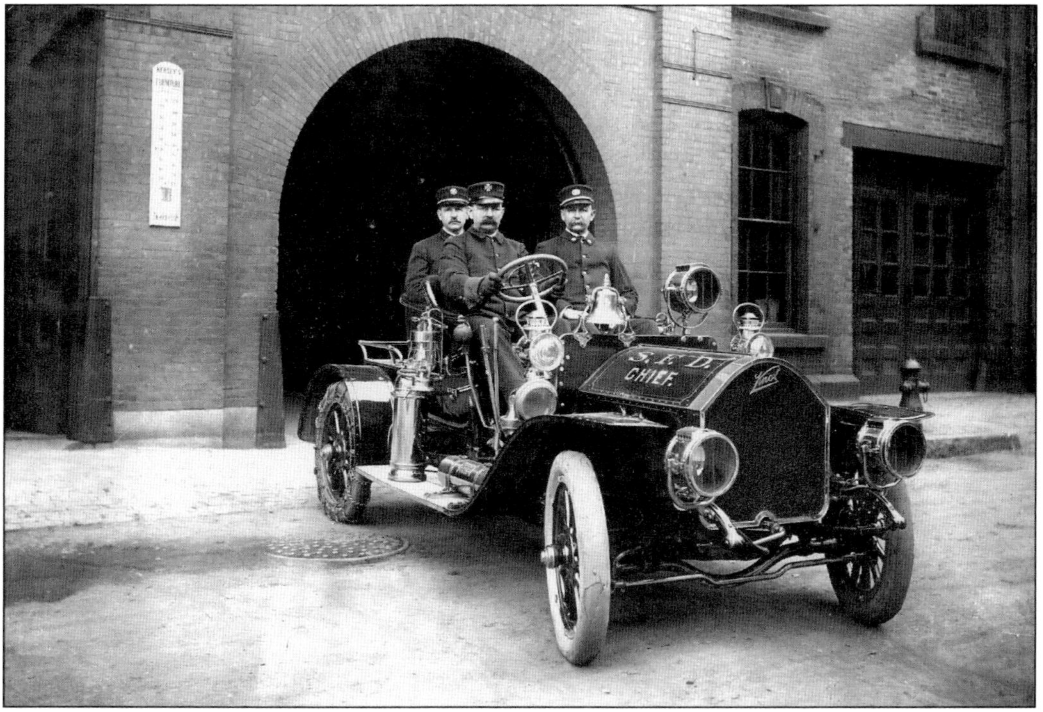

This is a view of Chief W. J. Littlefield with his driver and aid riding in one of the 1906 Knox automobiles. (Courtesy Noah Pooler family archives.)

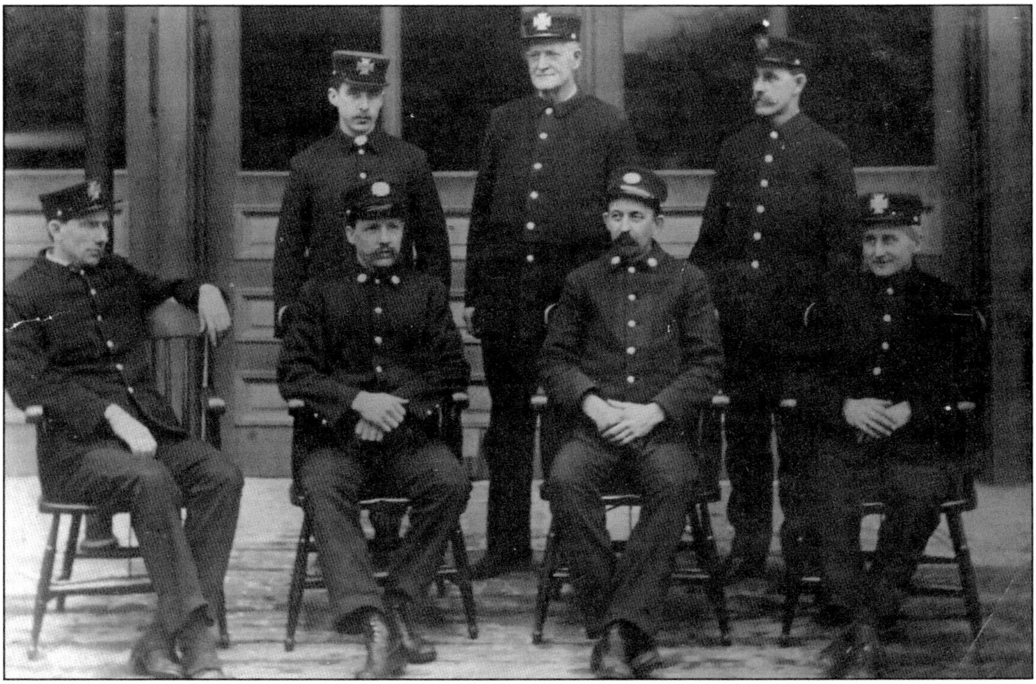

The men of Engine Company No. 5 at the Indian Orchard fire station are shown in this 1905 view. (Courtesy Noah Pooler family archives.)

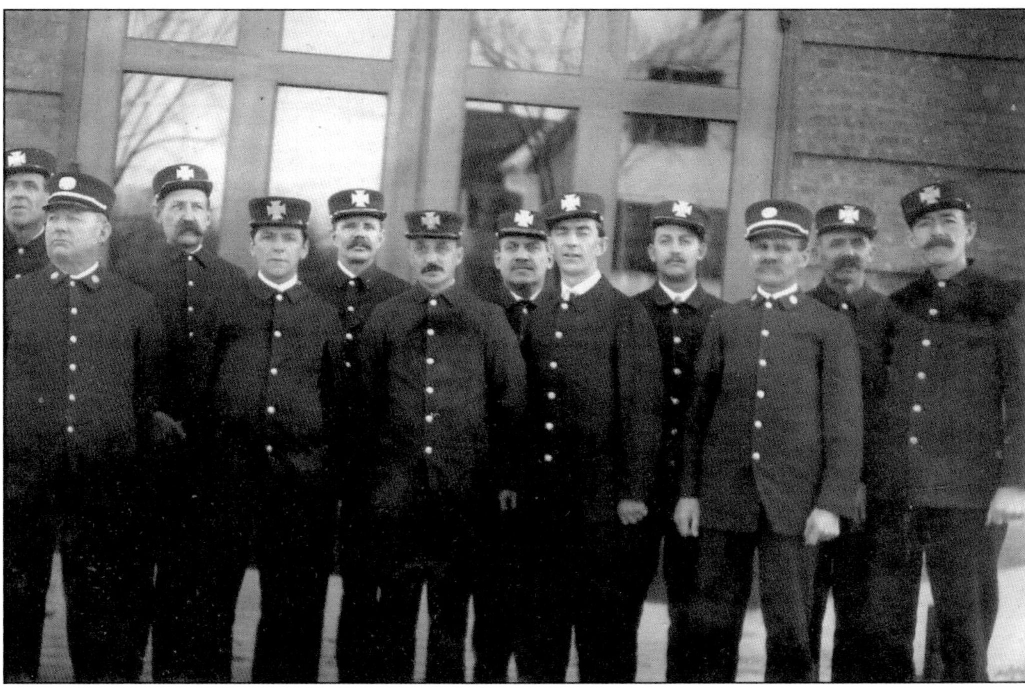

This 1905 view shows the firemen of Engine Company No. 4 on Walnut Street. The tall gentleman in the rear is Mr. Burton, who was the oldest firefighter on the department at that time. (Courtesy Noah Pooler family archives.)

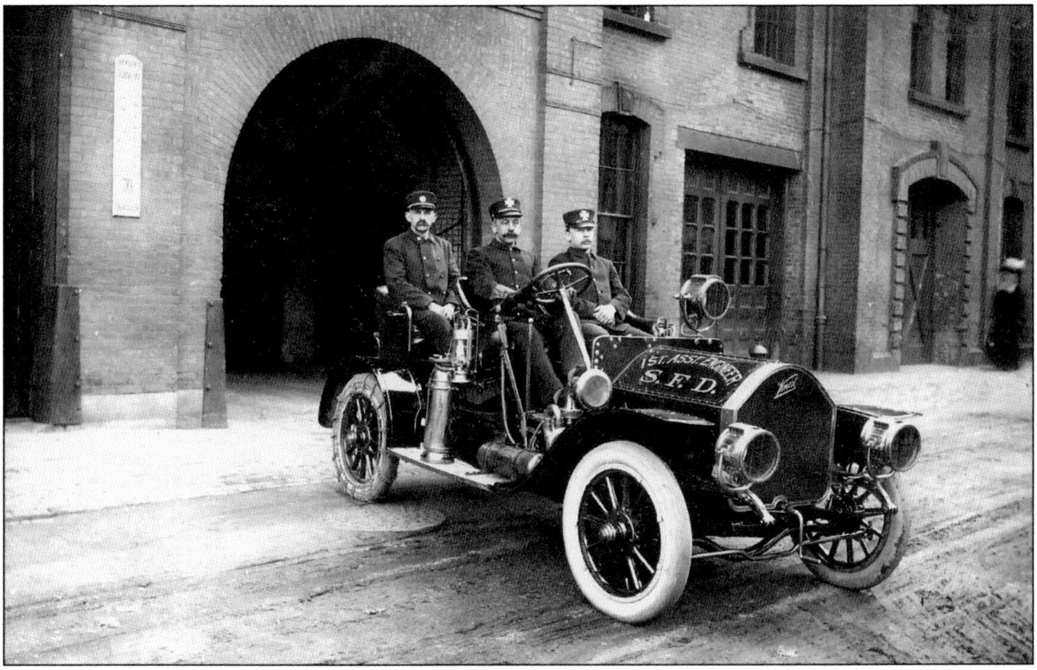

First assistant chief E. A. Kimball is shown here with his driver and aid on a 1906 Knox automobile in front of the Pynchon Street fire headquarters. (Courtesy Noah Pooler family archives.)

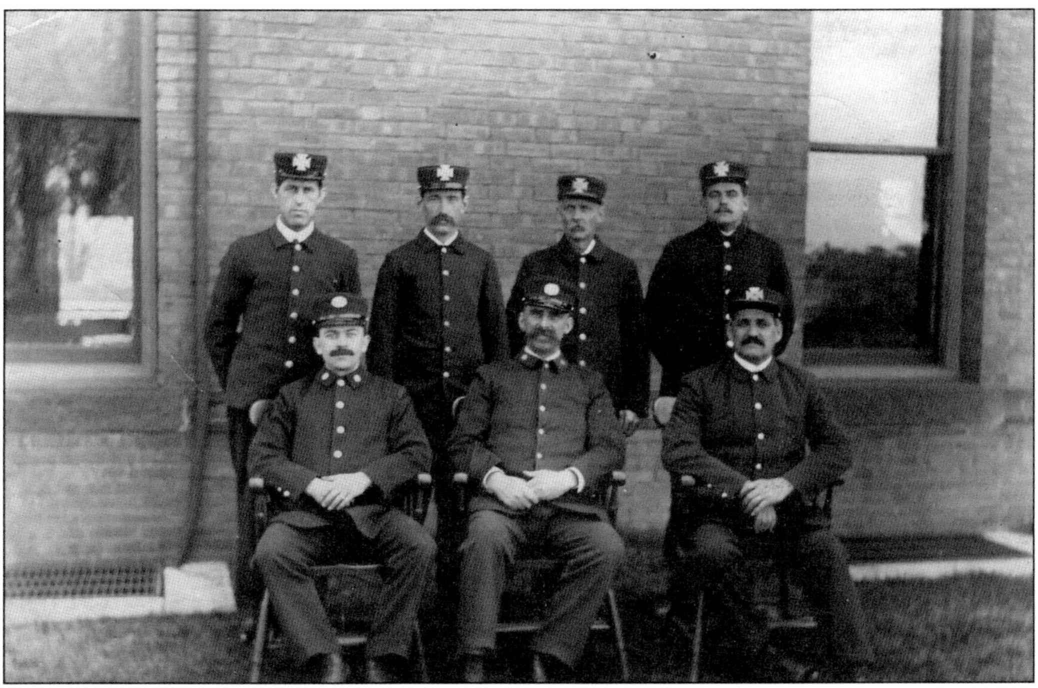

The men of Hose Company No. 13 at the Plainfield Street fire station are shown in this 1905 view. (Courtesy Noah Pooler family archives.)

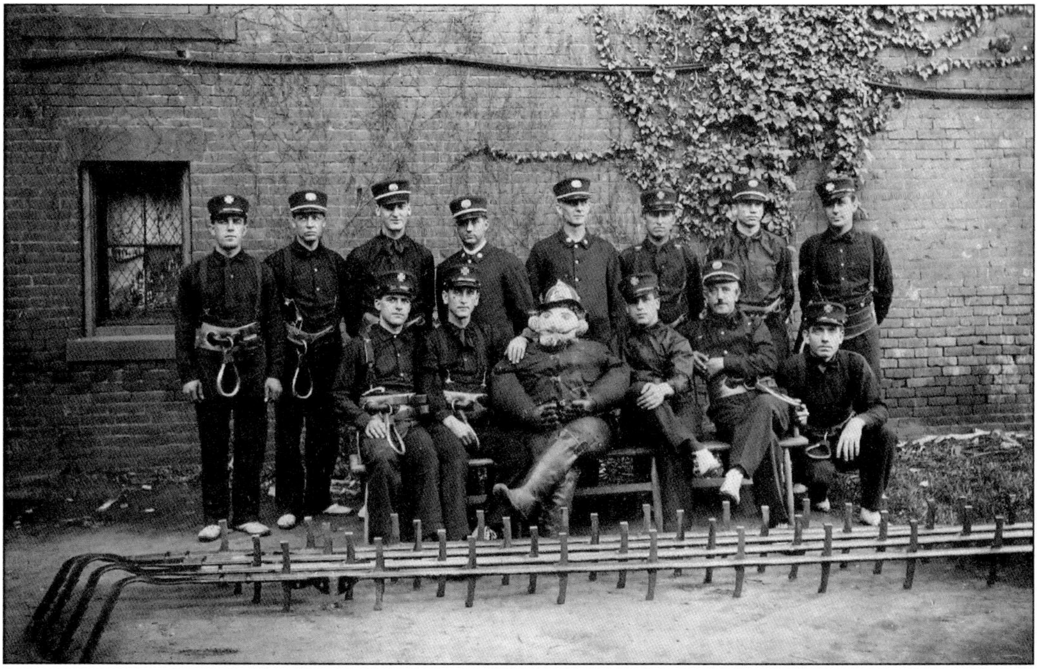

This photograph, taken on Labor Day in 1909, shows the students and instructors of the Springfield Fire Department Training School. In the second row on the left is Noah Pooler while the drillmaster Captain Strong is on the right. (Courtesy Noah Pooler family archives.)

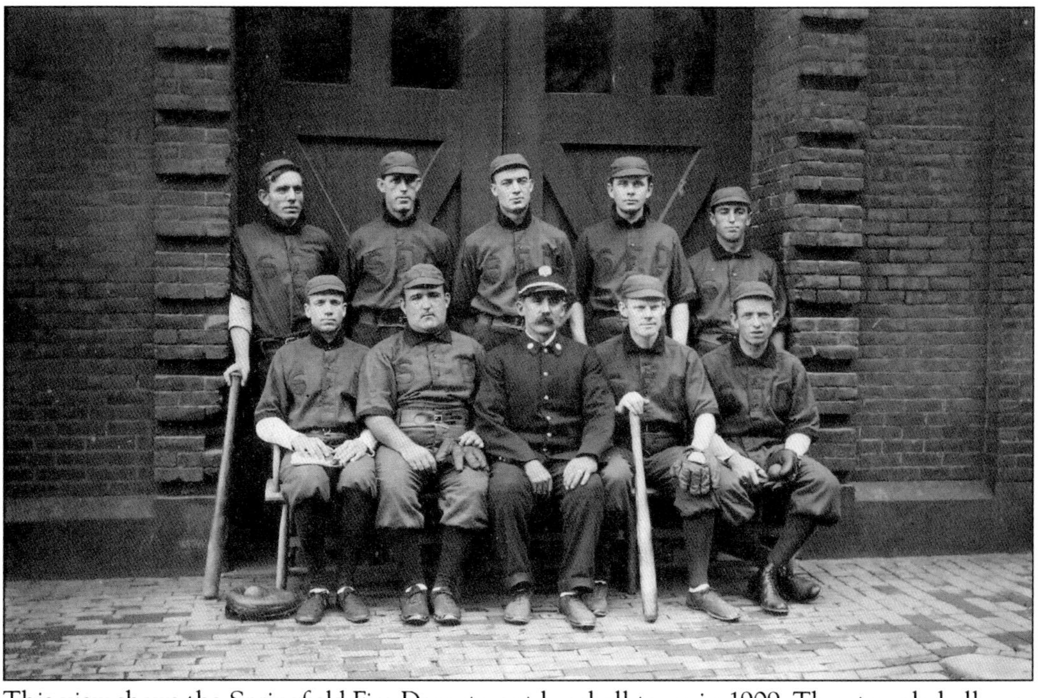

This view shows the Springfield Fire Department baseball team in 1909. They traveled all over New England playing against other fire department teams such as Holyoke, Hartford, Worcester, Providence, and Boston. (Courtesy Noah Pooler family archives.)

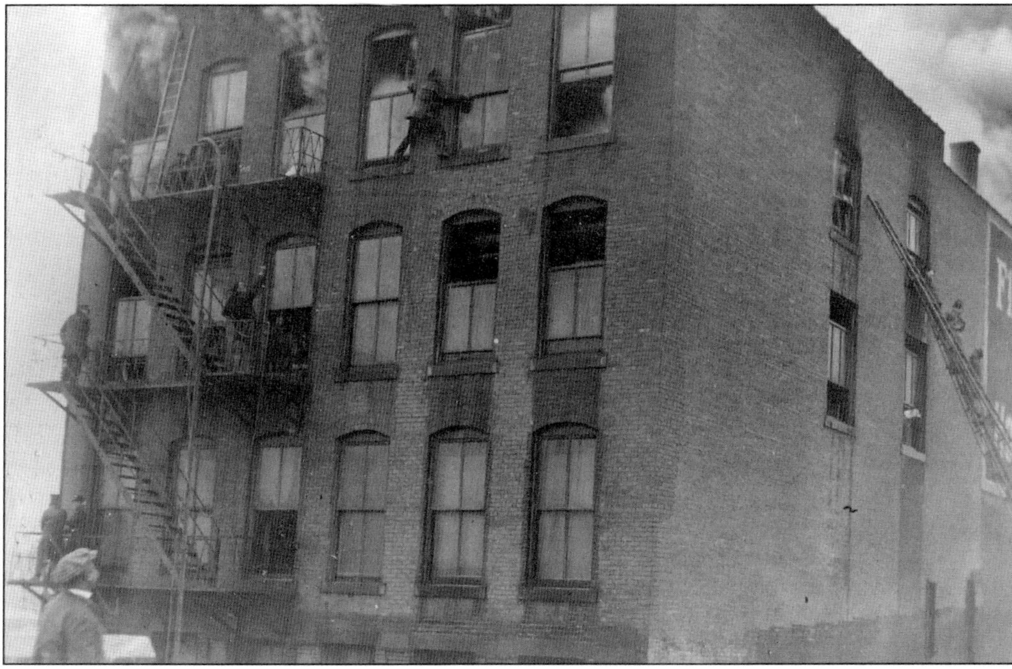

Springfield's firefighters regularly put their lives on the line in their performance of duty, as seen in this photograph taken at the Regan Building fire in 1915. (Courtesy Noah Pooler family archives.)

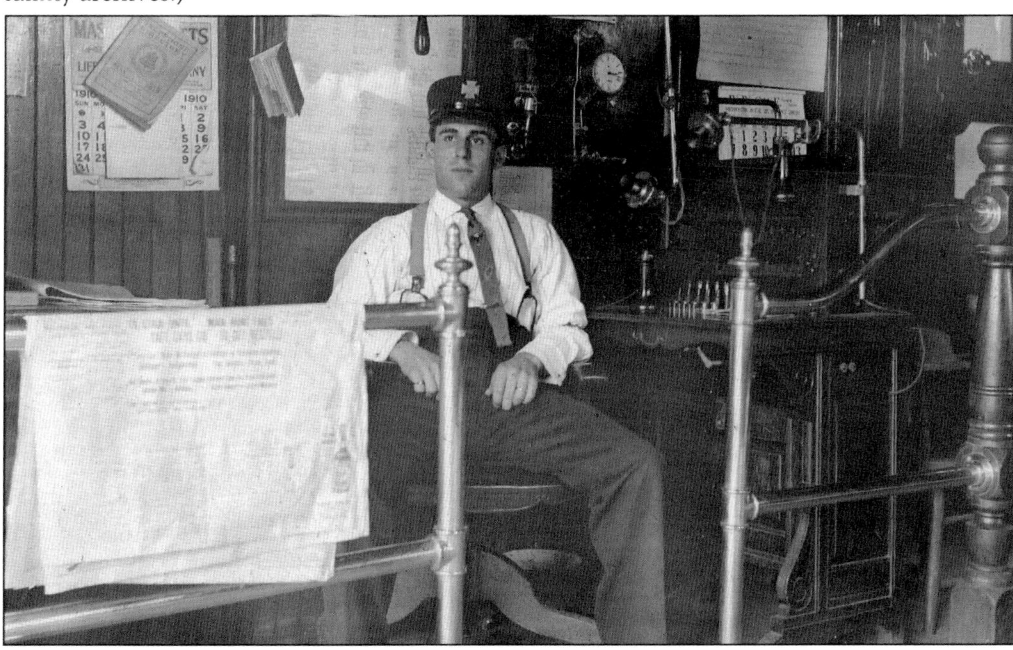

This is the watch desk and telephone switchboard at the Pynchon Street fire headquarters in 1910. Regulations called for all firemen to wear a necktie while on duty. This rule was not rescinded until the 1970s. At the time when this photograph was taken, almost all alarms came into the department on the fire alarm telegraph from one of the 262 red street boxes located around the city. (Courtesy Noah Pooler family archives.)

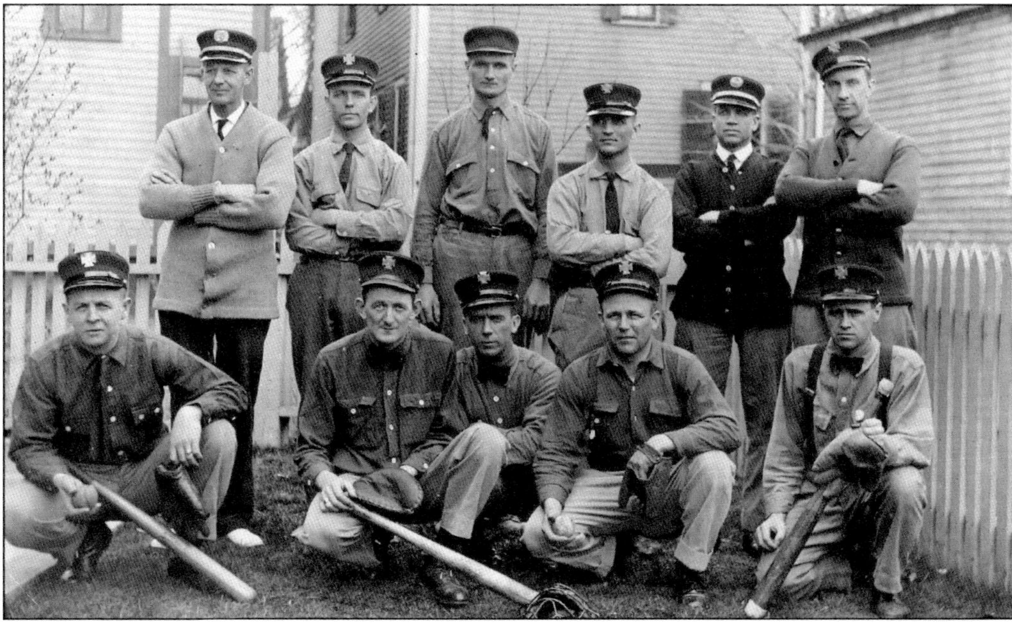

This view shows the Springfield Fire Department baseball team in 1914. (Courtesy Noah Pooler family archives.)

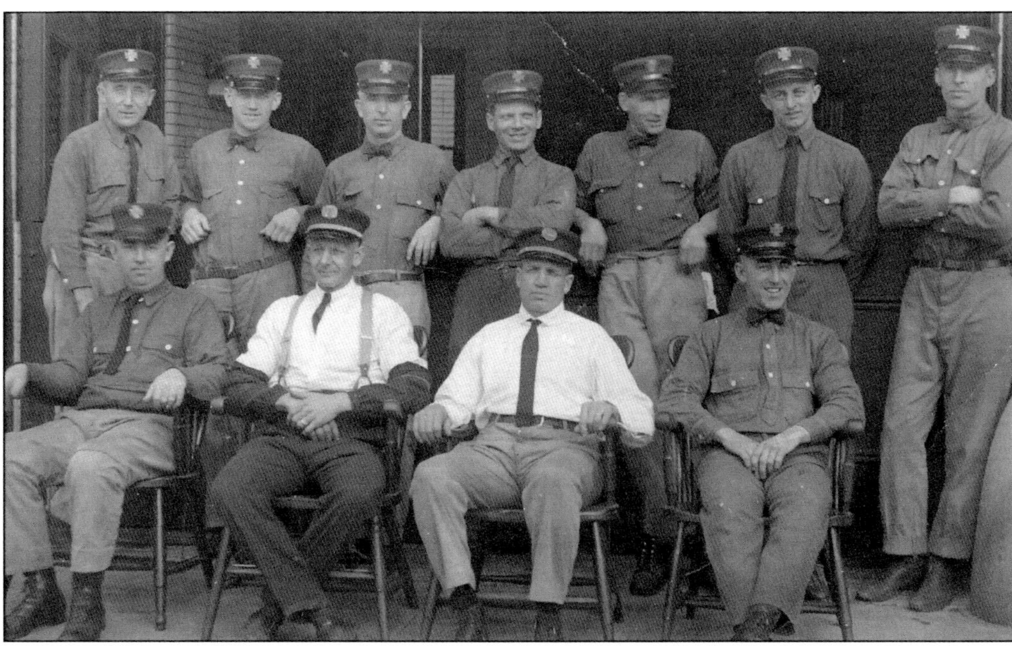

When this picture was taken in 1914, Springfield had less than 500 fire calls per year so many days could go by without a fire call. With the horses retired, the firemen had fewer duties to perform, although they still had to keep all the brass and nickel polished on the fire apparatus and make sure the station was clean and tidy. In the early 20th century, training was not as rigorous as it is today. (Courtesy Noah Pooler family archives.)

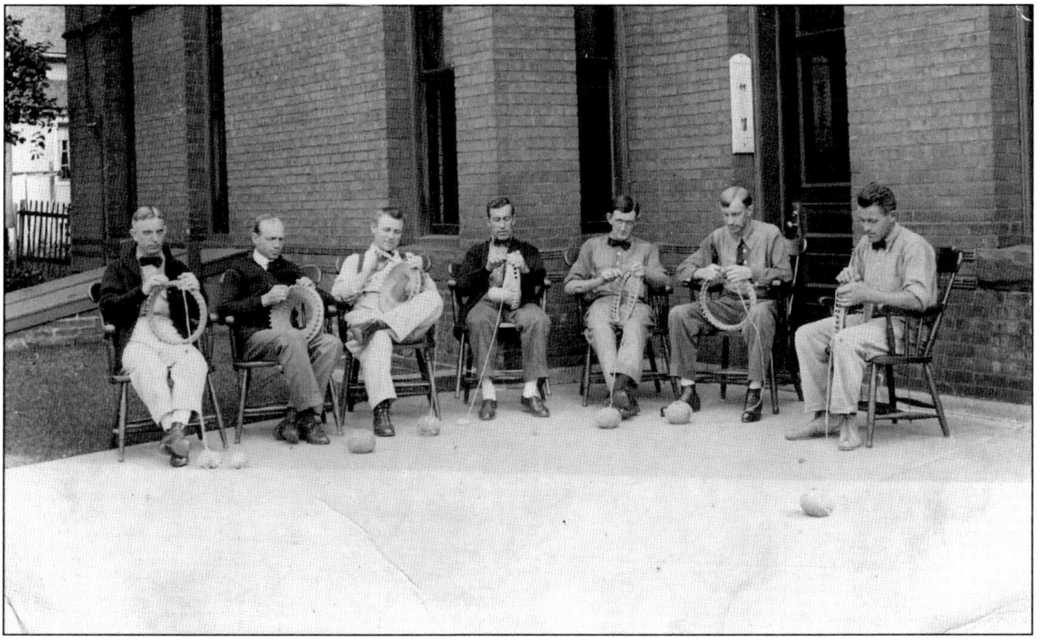

This 1918 view shows some of the firemen knitting woolen scarves for the military troops that were sent to France during World War I. (Courtesy Noah Pooler family archives.)

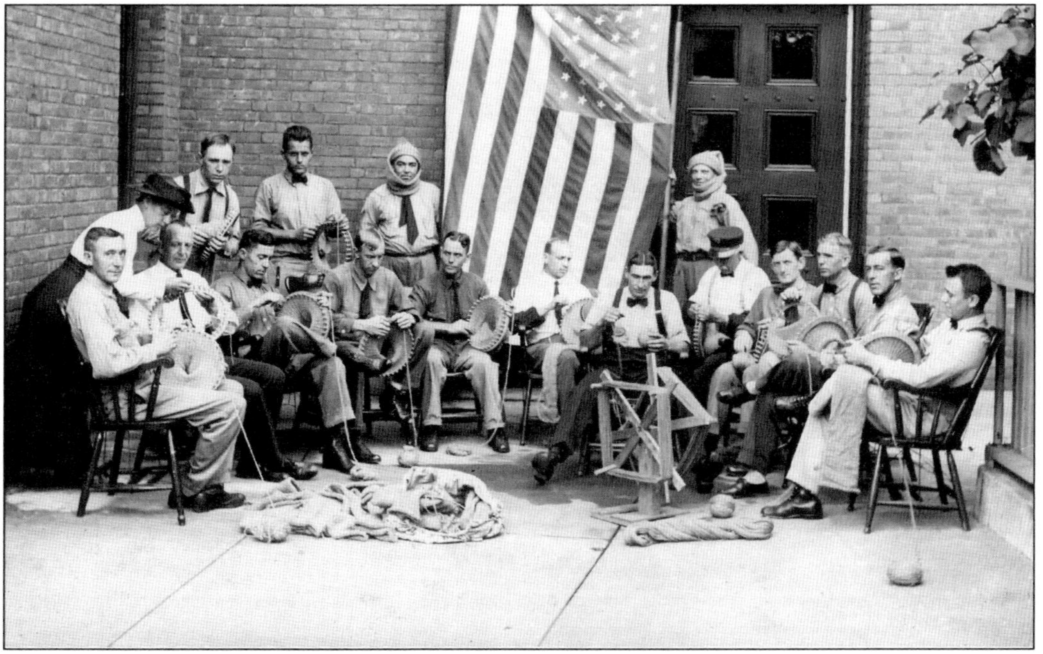

More Springfield firemen are doing their part for the war effort. The round objects they are using are yarn holders. (Courtesy Noah Pooler family archives.)

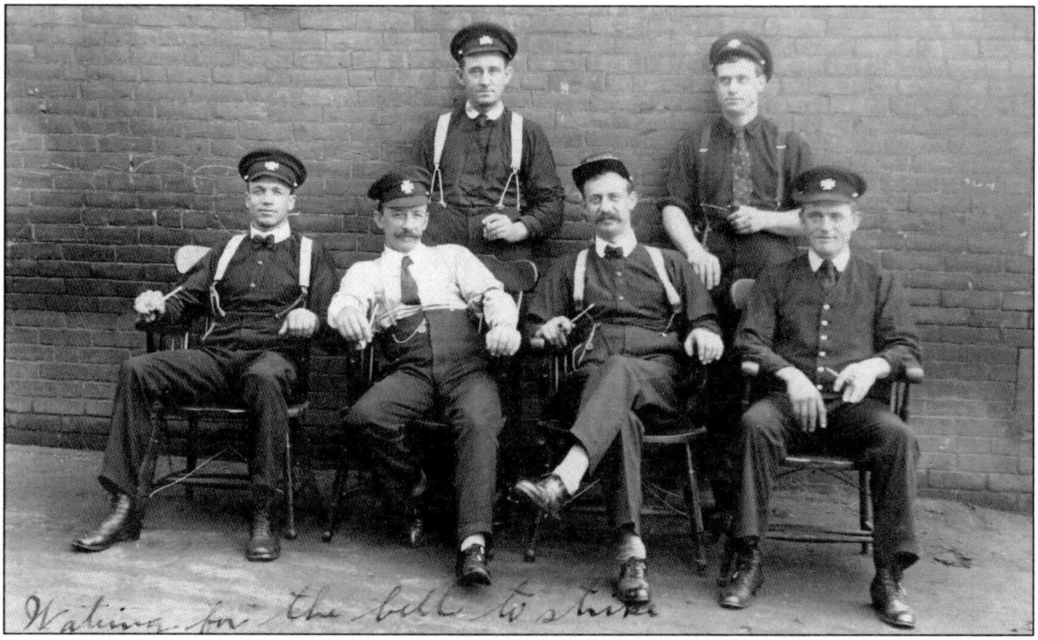

The title of this picture is *Waiting for the Bell to Ring.* Before the advent of radios and when telephones were still a novelty, the companies were dispatched by the fire alarm telegraph system that used bells to transmit the firebox number nearest the fire. Firefighters in the early 1900s were ever alert to the ringing of the fire bell. (Courtesy Noah Pooler family archives.)

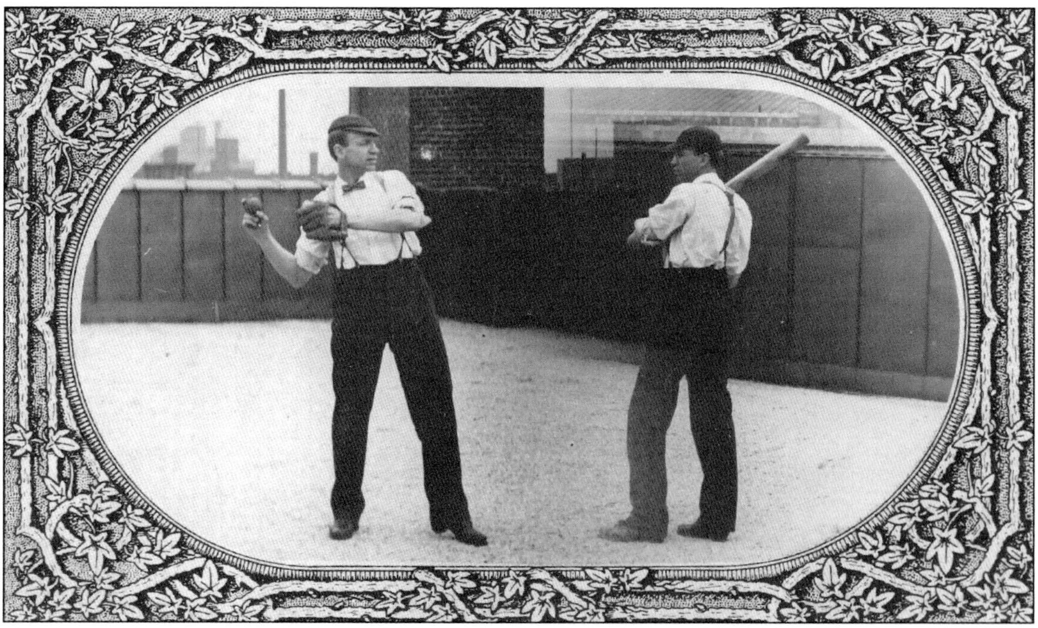

Noah Pooler was an avid photographer and took this unusual picture in 1914 throwing the ball to himself. (Courtesy Noah Pooler family archives.)

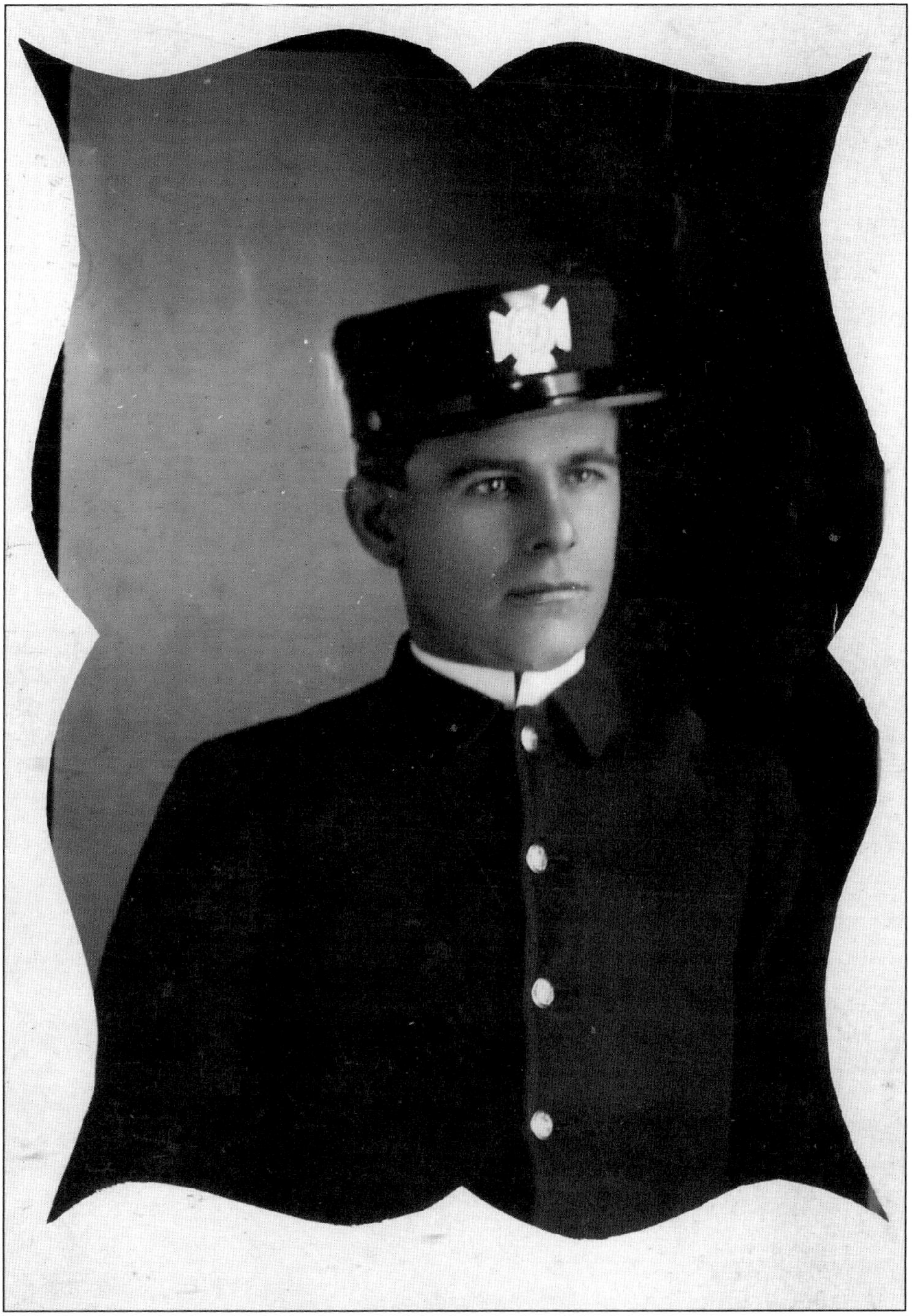

An unidentified Springfield firefighter in his dress uniform is shown in 1914. The firemen were required to wear their dress uniforms to work and, like the police and postal carriers, they were allowed to ride free on the Springfield Street Railways cars if they were in uniform. At the beginning and end of the shift they had to stand for inspection in their dress uniforms because the fire service has always been a paramilitary organization. (Courtesy Noah Pooler family archives.)

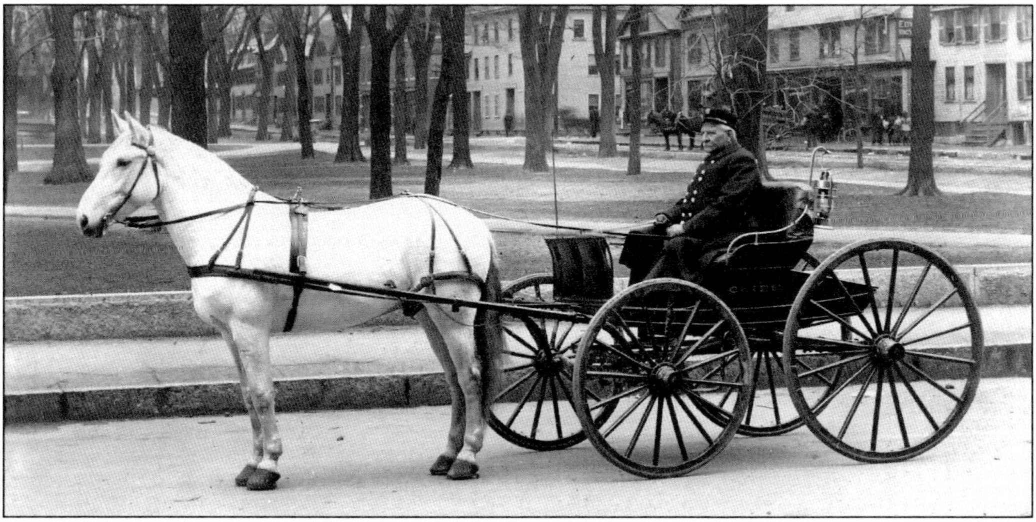

Springfield fire chief William H. Daggett is shown with his buggy before the advent of the automobile. In the days of the hand-drawn engines, the chief usually ran to the fire on foot or on some occasions rode a horse. (Courtesy Noah Pooler family archives.)

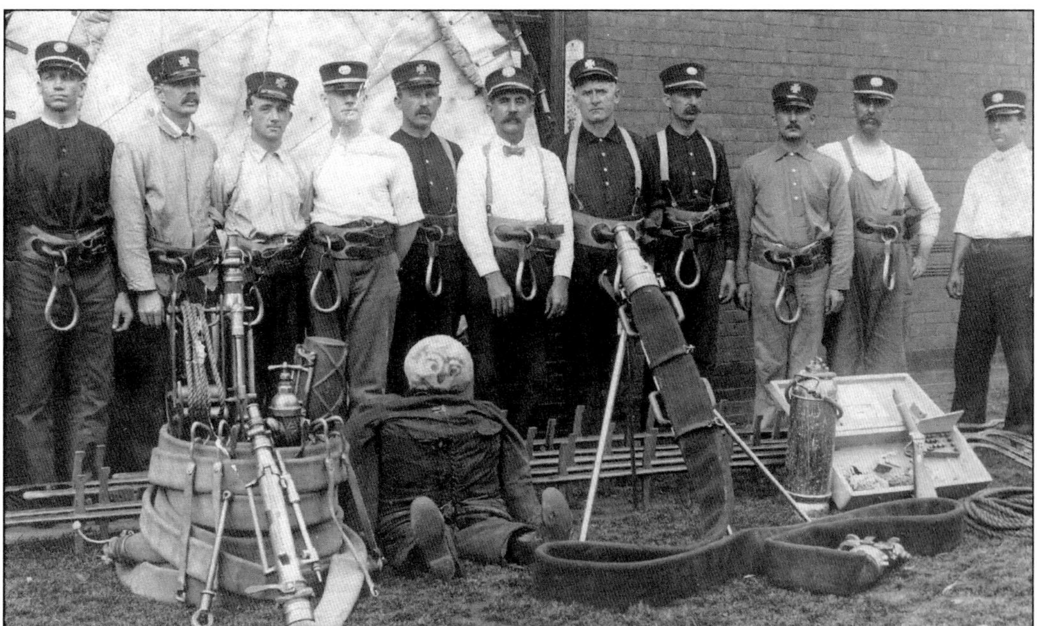

This view shows the training school classes with some of the equipment used by the cadets. (Courtesy Noah Pooler family archives.)

Firemen led a rigorous life earning less than $100 a month. In the horse-drawn era they had one day off in 30. They were also allowed two or three hours per day to go home for meals, as many firemen lived near the station. By 1914, when this photograph was taken, the men had one day off in eight. Thanks to the labor union movement, the work week gradually dropped first to 84 hours, then to 70 hours, and finally in 1948 to a 48 hours. In order to institute the 48-hour workweek, the department hired 138 new men in February 1948. The firefighters' labor unions fought hard over the years to bring about the work schedules and staffing that paid firefighters in the United States and Canada have today. (Courtesy Noah Pooler family archives.)

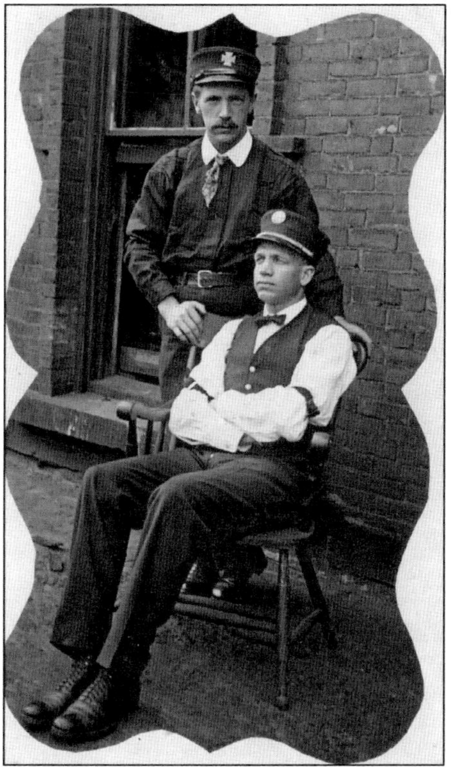

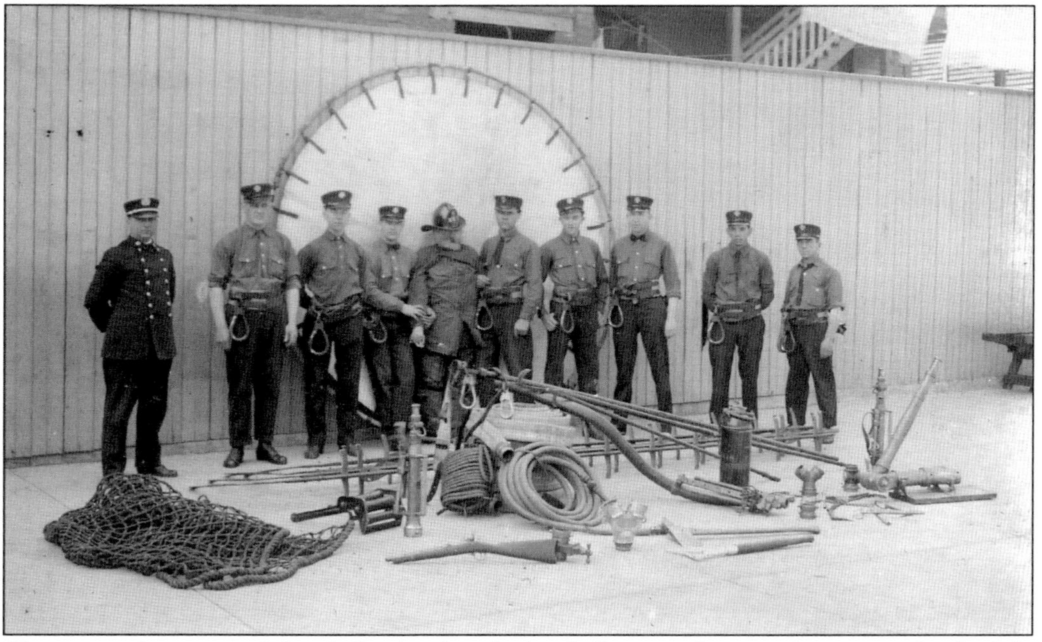

This is another view of the drill school with the drillmaster Captain Strong on the far left. (Courtesy Noah Pooler family archives.)

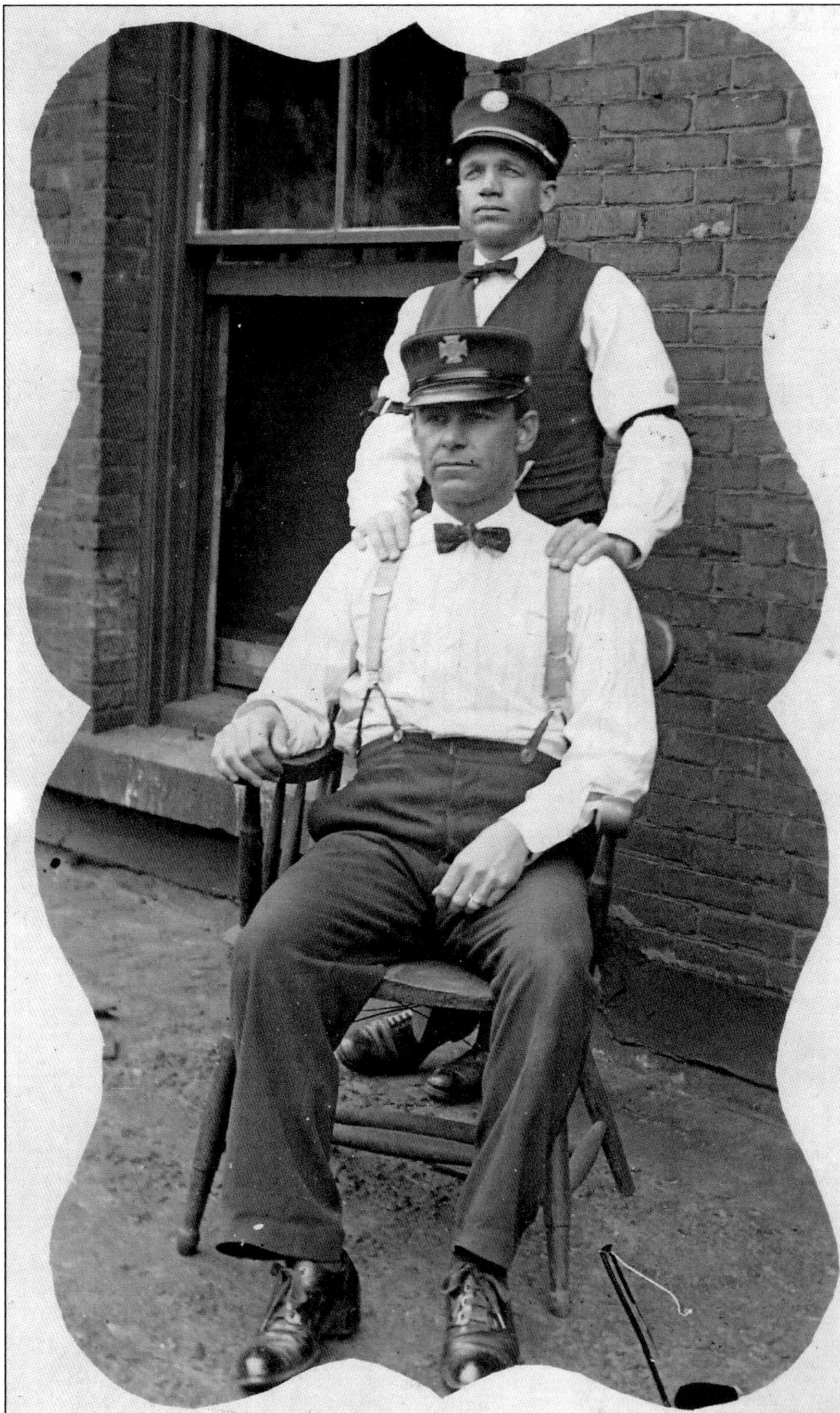

Noah Pooler and an unidentified firefighter are shown in this 1914 scene. (Courtesy Noah Pooler family archives.)

Six

The Modern Era

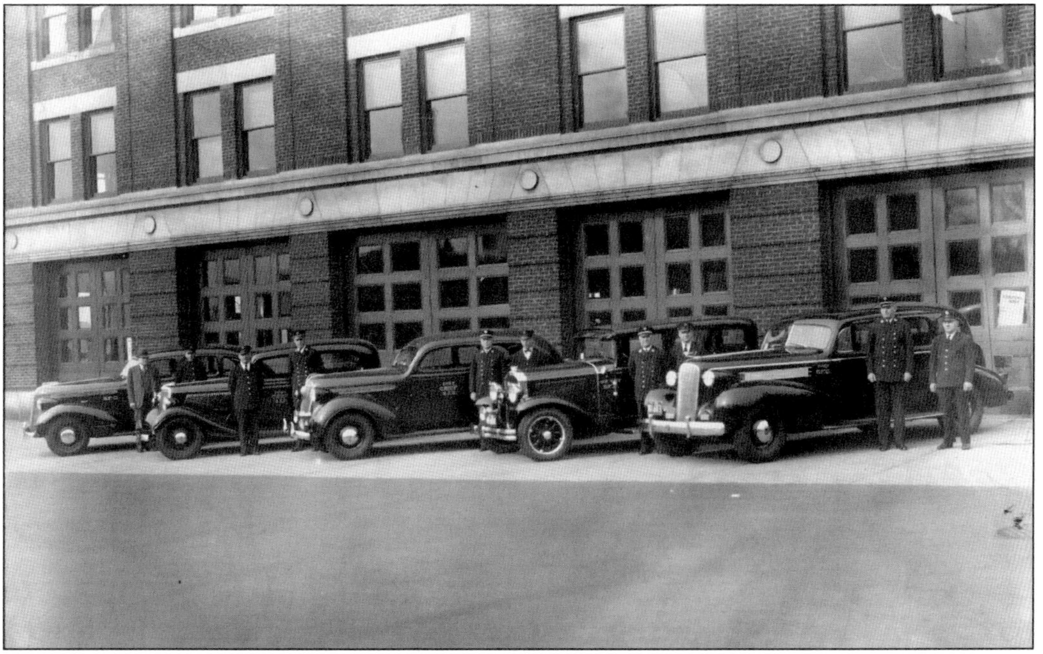

This photograph shows a lineup of all the fire department automobiles in front of the Court Street fire headquarters in the 1940s. (Courtesy Raymond Pond.)

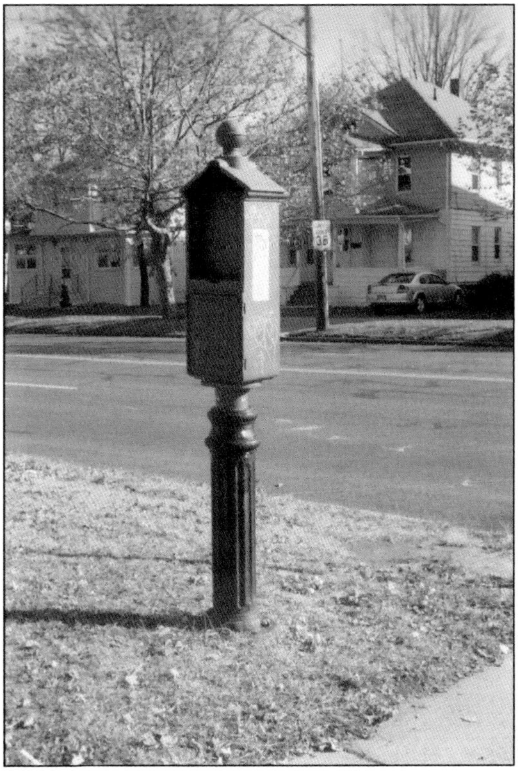

At one time, Springfield had over 350 fireboxes located at street corners and inside important city buildings. In 1868, the first fire alarm system was installed and was originally considered a useless contrivance and a waste of taxpayers' money. In 1885, the city created the post of fire alarm superintendent and upgraded the system to the more dependable Gamewell fireboxes. By 2006, most of the fireboxes had been removed as the city moved toward no longer having a public fire alarm system. (Courtesy Connecticut Fire Museum archives.)

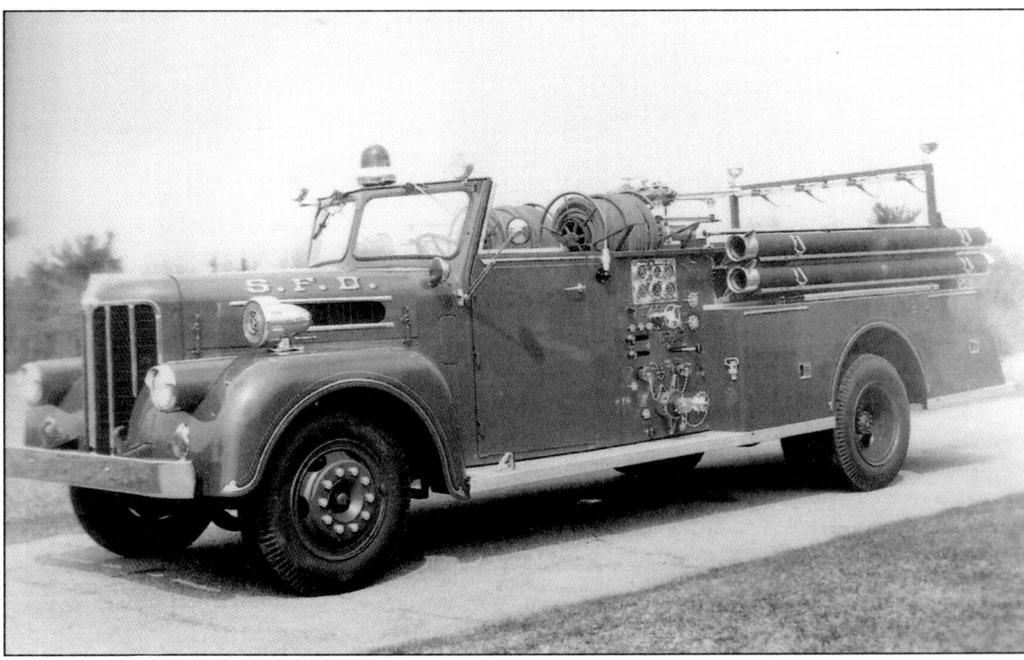

In the 1950s, Springfield began to purchase apparatus from Maxim Motor Company in Middleboro to replace many of the Seagrave and Mack pumpers and hose wagons purchased in the 1930s. (Courtesy Connecticut Fire Museum archives.)

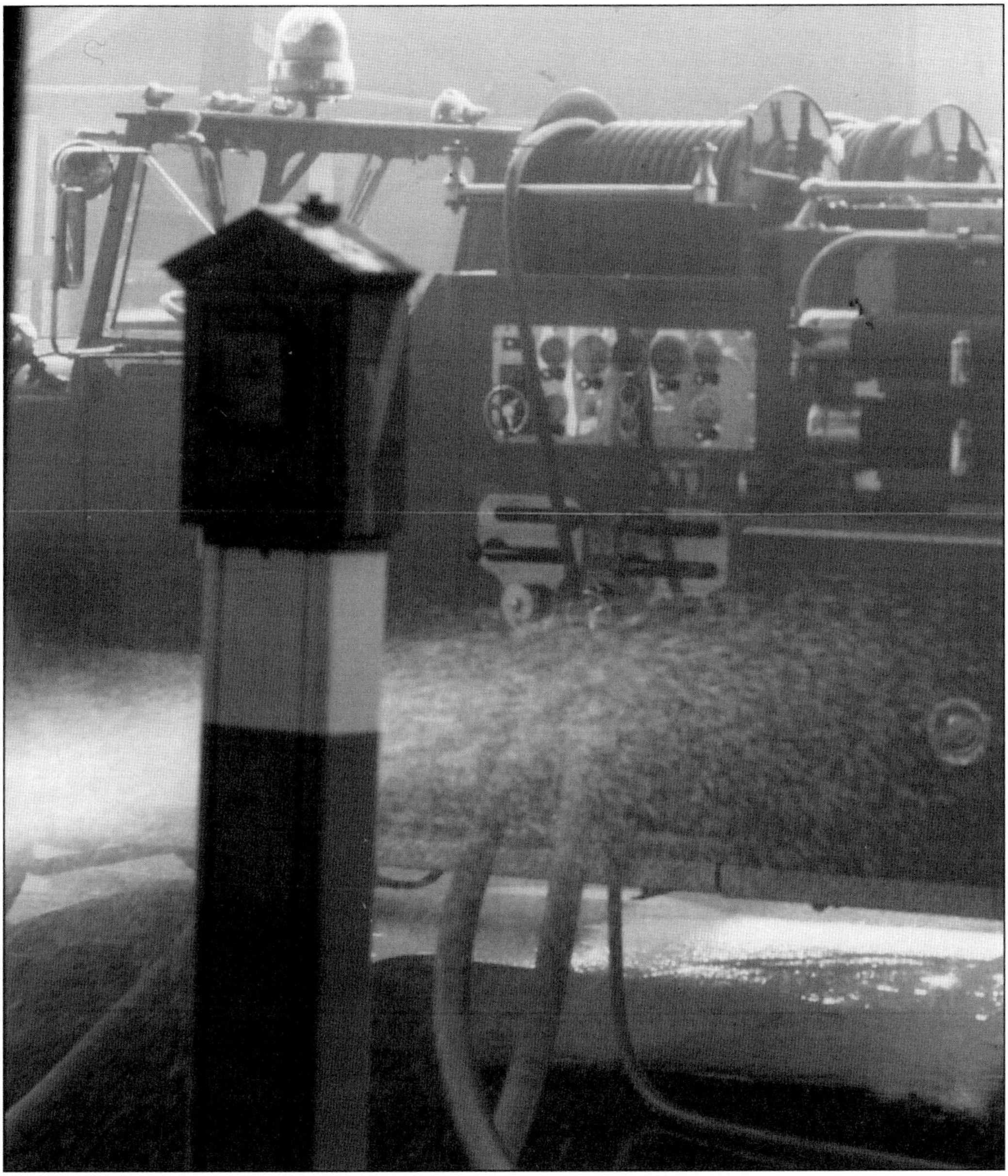

This view shows one of the Maxim pumpers in action at a fire in the 1970s. The alarm may have come in from the firebox in the foreground. In 1955, the National Board of Fire Underwriters recommended that the city increase the number of boxes from 350 to 700 due to new building construction and population increase. Problems with false alarms from fireboxes began in the 1950s and have continued to increase. (Courtesy Geoffery Neilsen.)

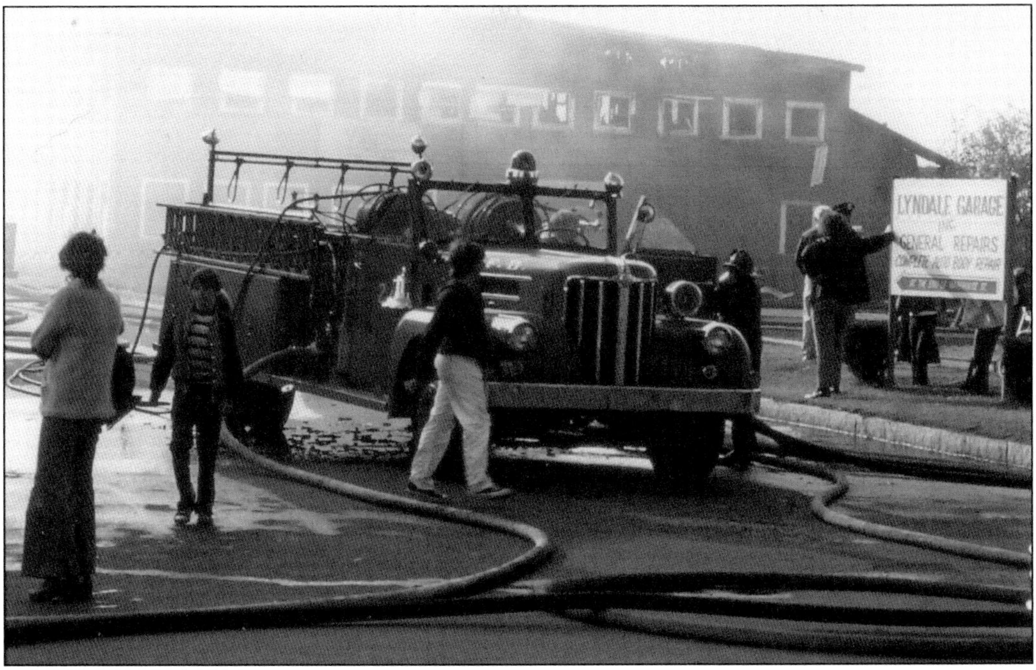

Another Maxim pumper is in action at a building fire in the 1970s. Today the public is not allowed near the equipment for safety reasons. The 2.5-inch hose in the foreground is charged with over 150 pounds of pressure and could cause an injury if it broke. (Courtesy Geoffery Neilsen.)

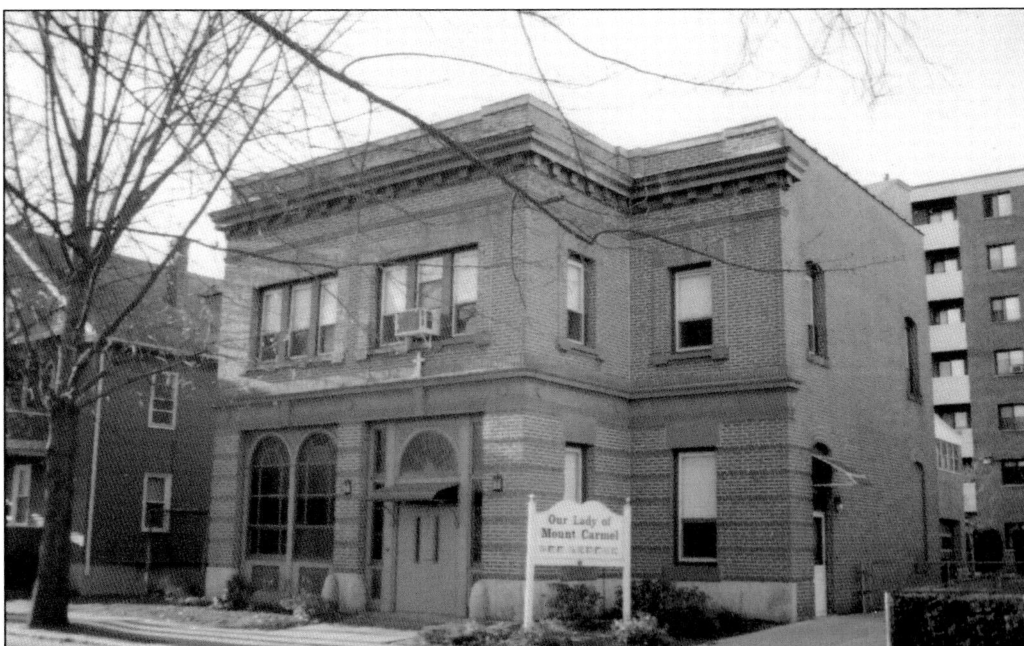

This 2006 view shows the Margaret Street fire station in its new role as a day care center for Our Lady of Mount Carmel. (Courtesy Connecticut Fire Museum archives.)

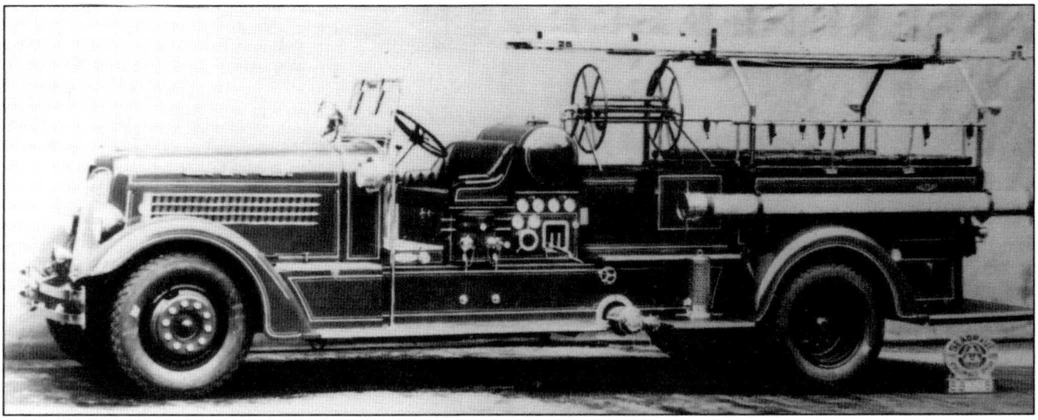

The 1937 Seagrave 1,200-gallon-per-minute pumper of Engine Company No. 4 was headquartered at the Walnut Street fire station shown in this photograph. This truck was sold to a collector in 1975. Compare this to engine 4 of 1914 shown on page 64.

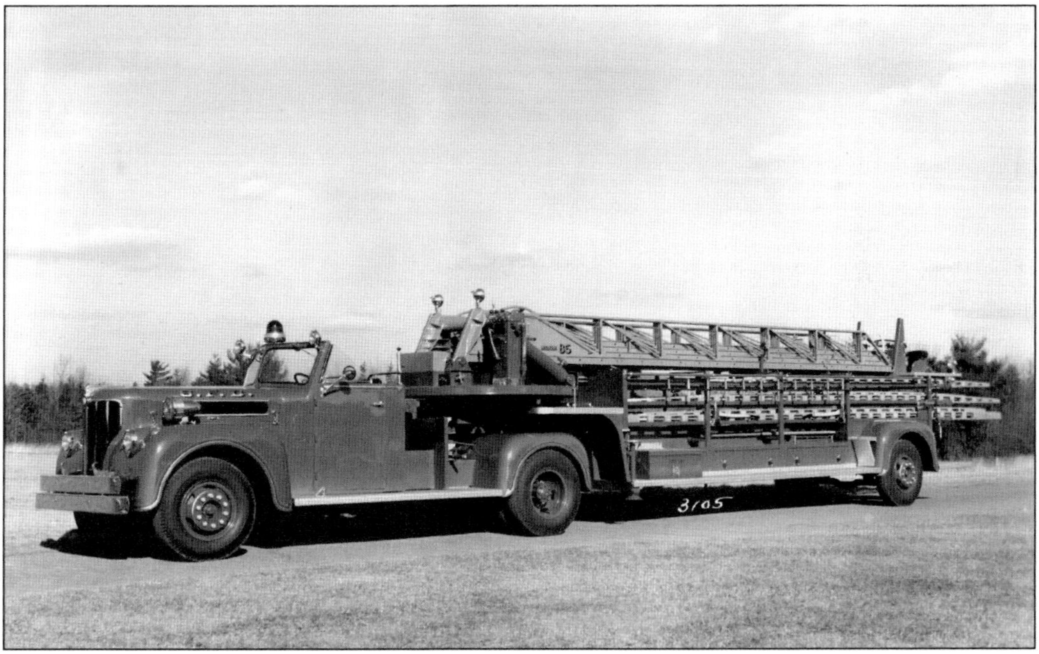

This is a builders photograph of the 1954 Maxim tillered 100-foot aerial ladder purchased by the city. In 1954, the Springfield Fire Department had 60 pieces of equipment, including fire apparatus, the chief's automobiles, and service trucks all maintained by the repair division located at 724 Sumner Avenue and managed by master mechanic Norman Robert. (Courtesy Raymond Pond.)

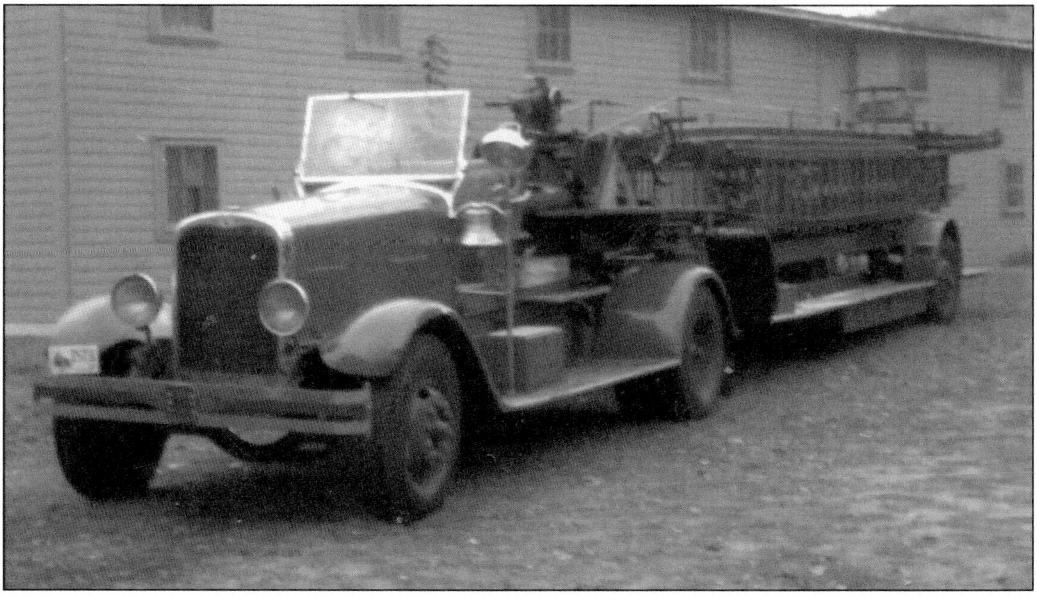

This is a view of Ladder Company No. 13 after it was purchased at auction by the Connecticut Fire Museum in 1973. The tractor is a 1934 Seagrave and the trailer is a 1929 Seagrave 85-foot aerial ladder. This tractor replaced a Knox tractor that originally pulled a converted horse-drawn ladder truck. In 1929, the horse-drawn unit was replaced with the new Seagrave trailer seen here. This is now on display at the museum in East Windsor, Connecticut. (Courtesy Connecticut Fire Museum archives.)

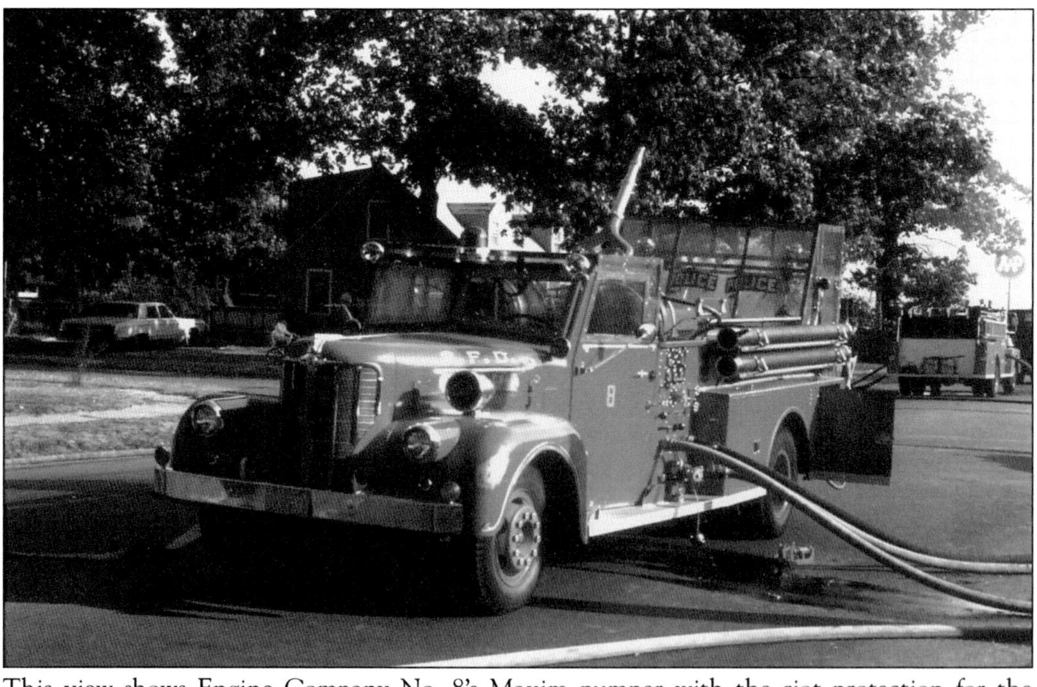

This view shows Engine Company No. 8's Maxim pumper with the riot protection for the firefighters that was put on the apparatus during the civil unrest of the late 1960s. (Courtesy Geoffery Neilsen.)

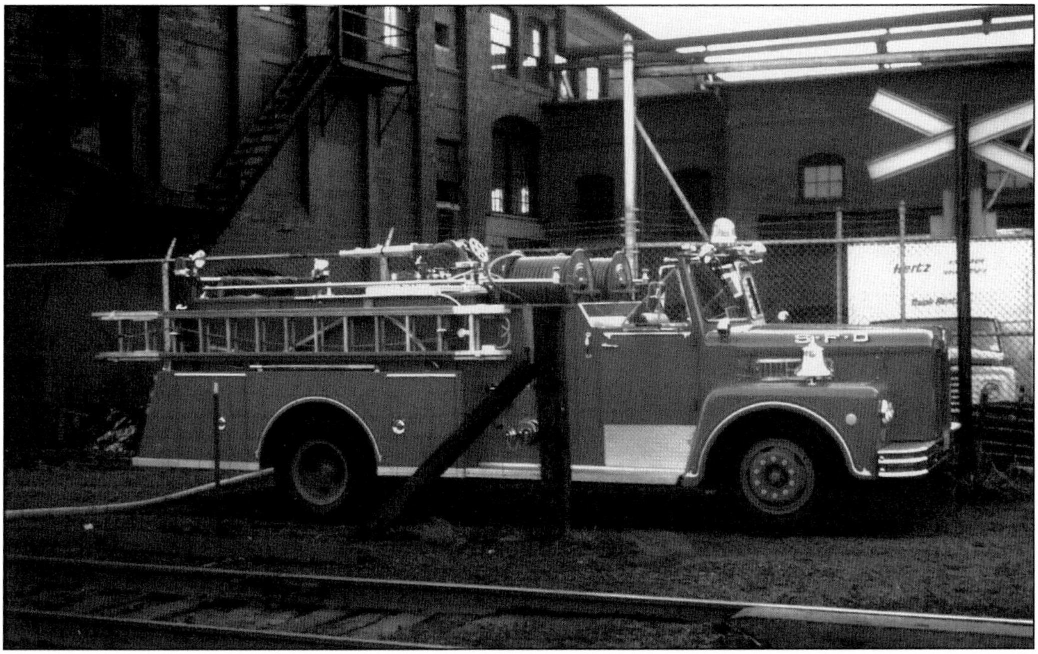

This Maxim pumper from the 1960s served the Springfield Fire Department. Maxim Motors built its first fire apparatus in 1914, and before the company closed in 1990, Maxim had a reputation for high quality and dependability. For a short time in the 1980s, the Seagrave Corporation, which also built Seagrave fire apparatus, owned Maxim. (Courtesy Geoffery Neilsen.)

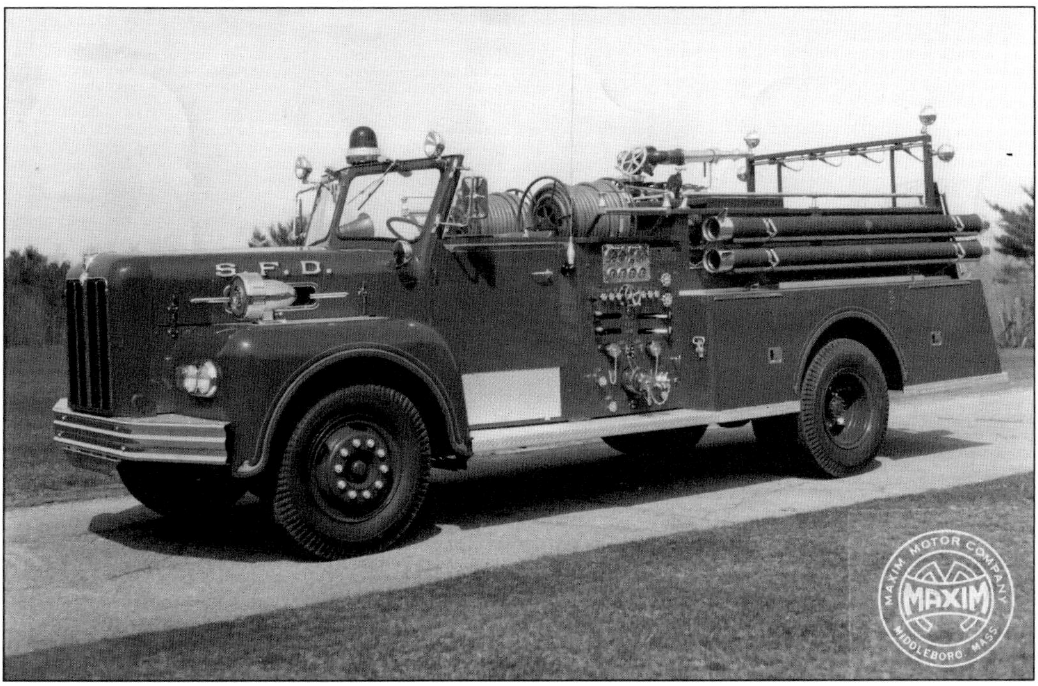

This is a builders photograph of a Maxim S model built for Springfield. The era of the open-cab truck would be over by the 1980s. In 1965, the department reported that it still had one 1916 and one 1930 pumper on the roster as spares. (Courtesy Raymond Pond.)

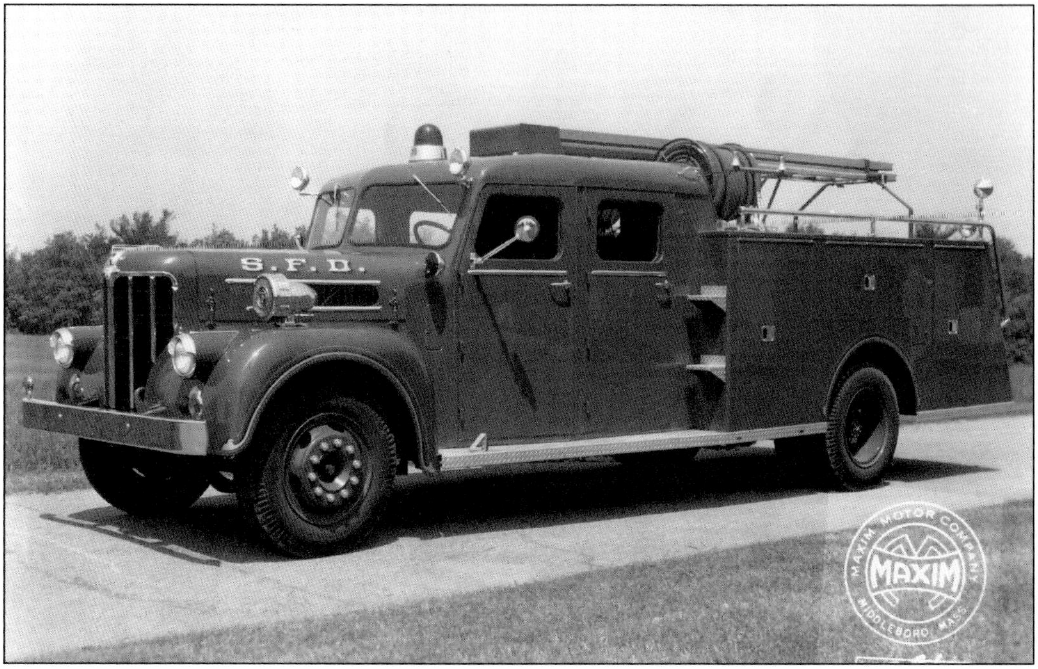

This photograph shows a 1953 Maxim F model rescue truck used by Squad Company B. Compare this to the 1911 Knox truck on the bottom of page 72. In 1973, another S model Maxim rescue unit would replace this truck. (Courtesy Raymond Pond.)

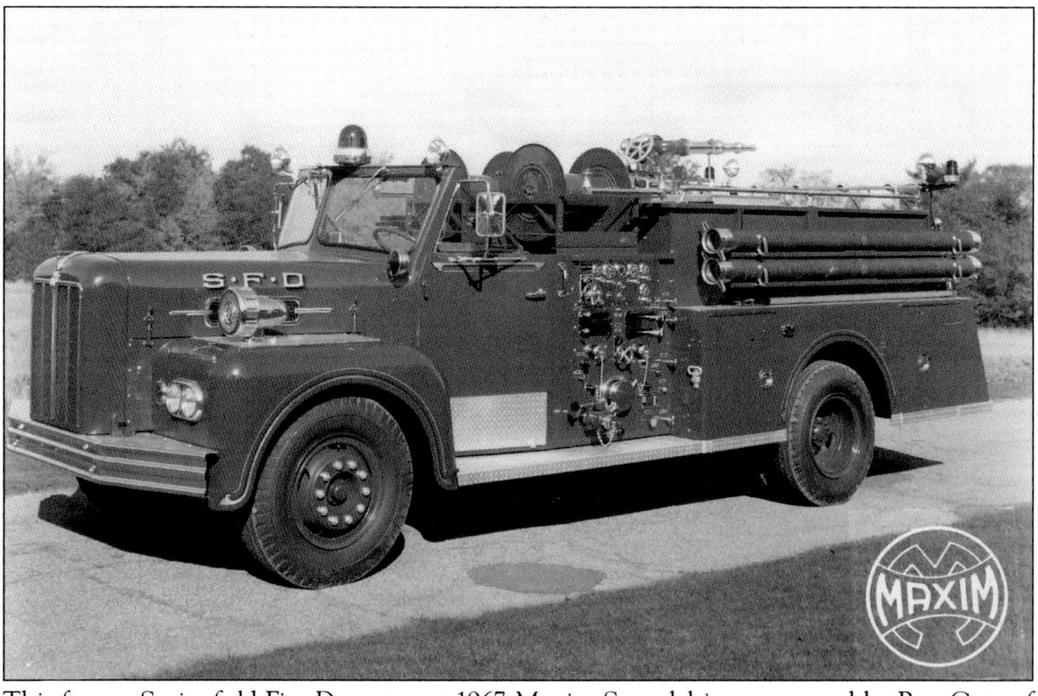

This former Springfield Fire Department 1967 Maxim S model is now owned by Ben Cote of East Long Meadow and can be seen at numerous firemen's parades in the summer months. (Courtesy Raymond Pond.)

This 2006 view shows the former North Main Street fire station in its new role as the office for the Pioneer Valley Transit Authority. (Courtesy Connecticut Fire Museum archives.)

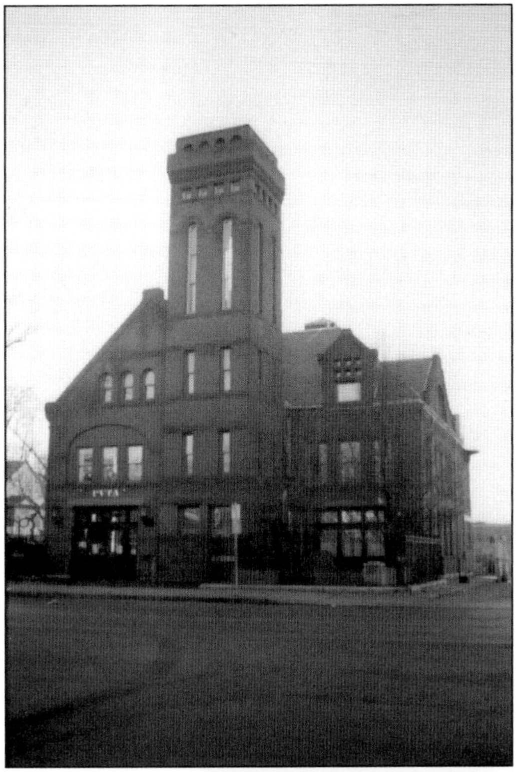

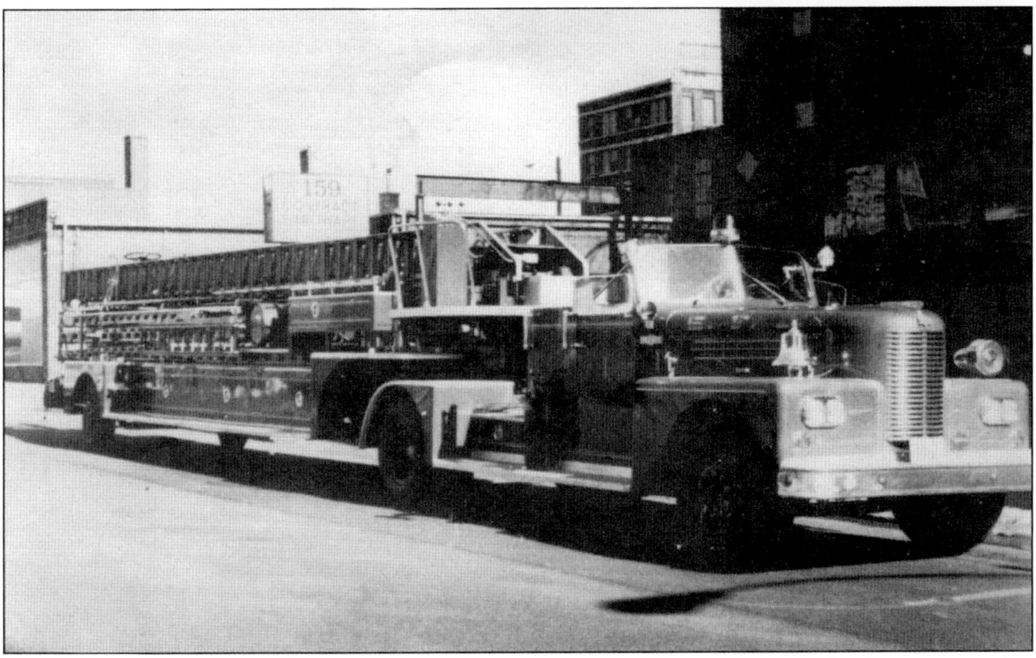

Ladder Company No. 1 operated this 1964 Pirsch 100-foot aerial ladder in the late 1960s. This was the only piece of Pirsch apparatus used by the department. Peter Pirsch and Sons was a family-owned company located in Kenosha, Wisconsin. Compare this to Ladder Company No. 1's first ladder truck on page 29. (Courtesy Alarm Room News.)

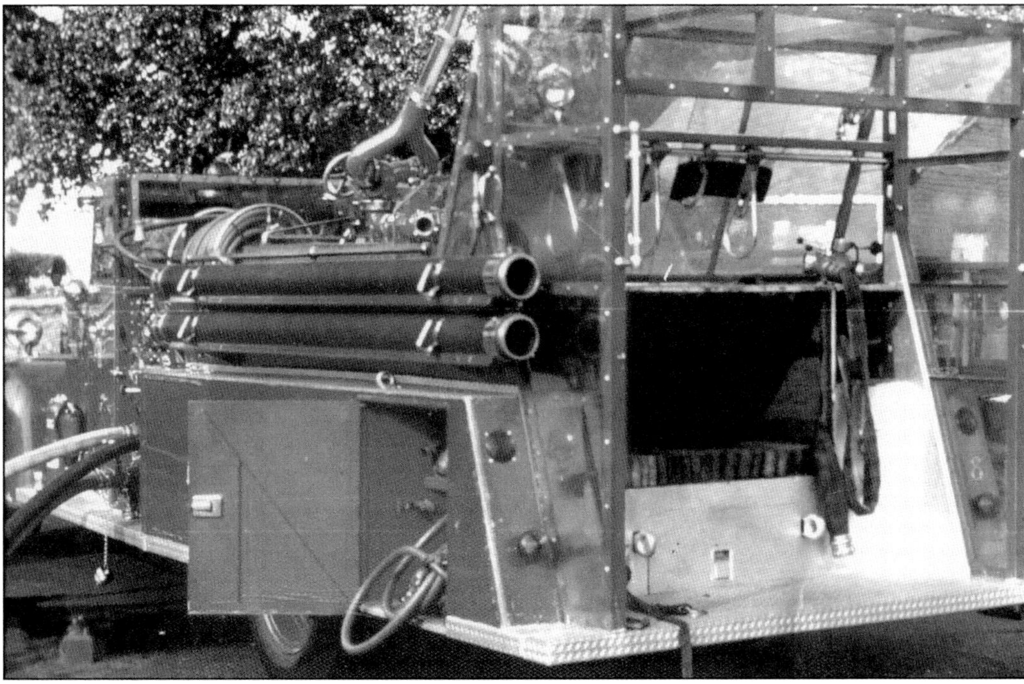

Here is a closer look at the riot gear that the Springfield Fire Department maintenance division designed and installed on many of the city apparatus during the time of civil unrest in Springfield. (Courtesy Geoffery Neilsen.)

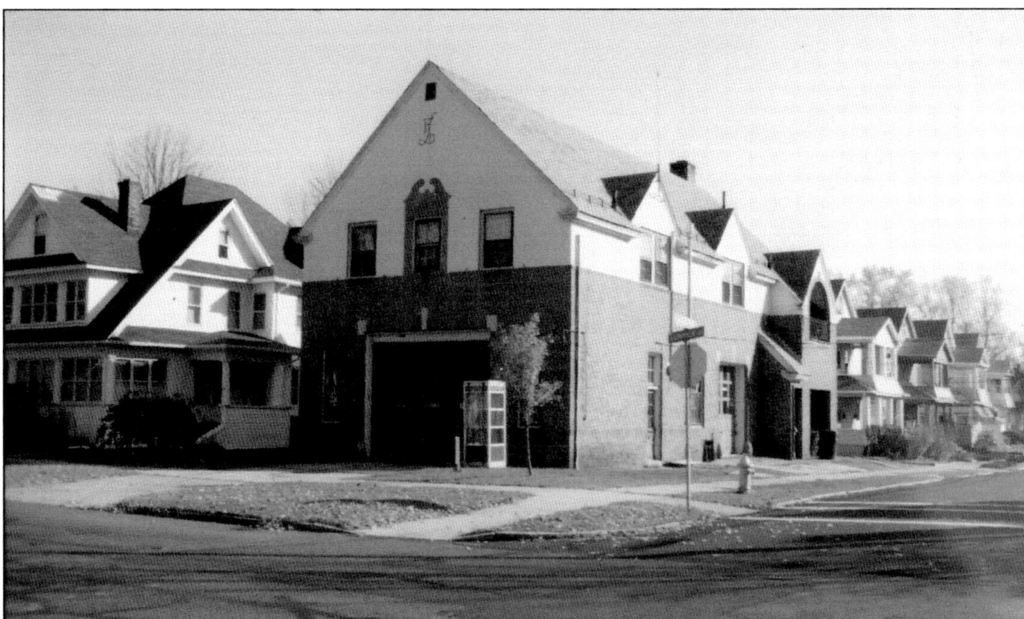

This is a view of the fire station at 1045 Sumner Avenue taken in 2006. This is the home of Engine Company No. 14, which was formed from Squad A back in the 1930s.

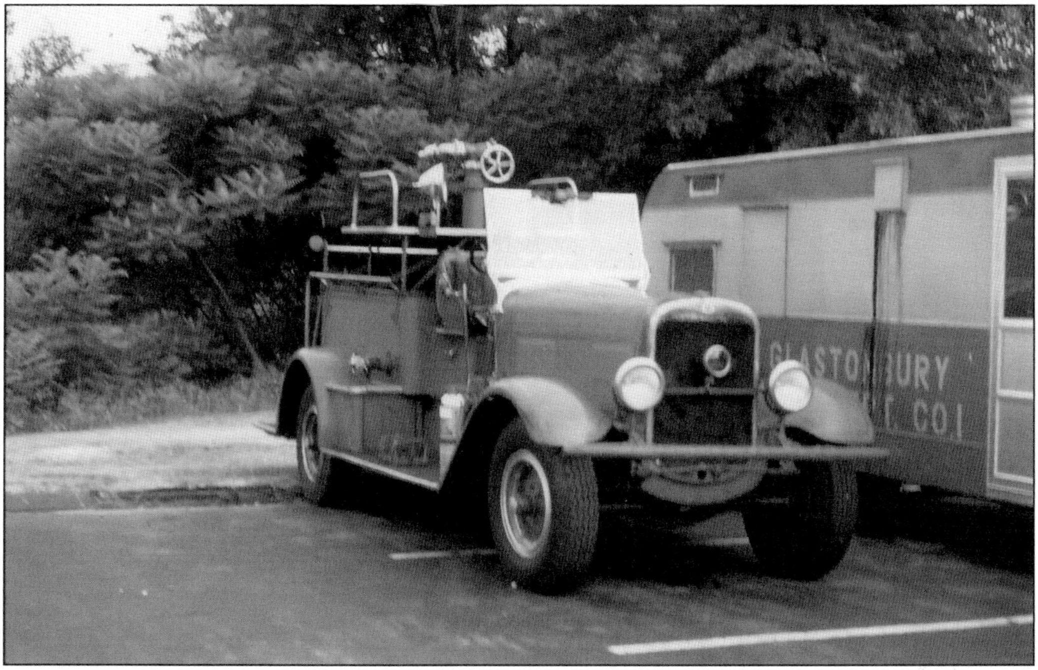

During the 1960s, the Glastonbury Fire Department in Connecticut had this former Springfield 1925 Seagrave hose wagon that it was going to convert into a deluge wagon. What happened to the project is unknown, and the truck was apparently scrapped. (Courtesy Geoffery Neilsen.)

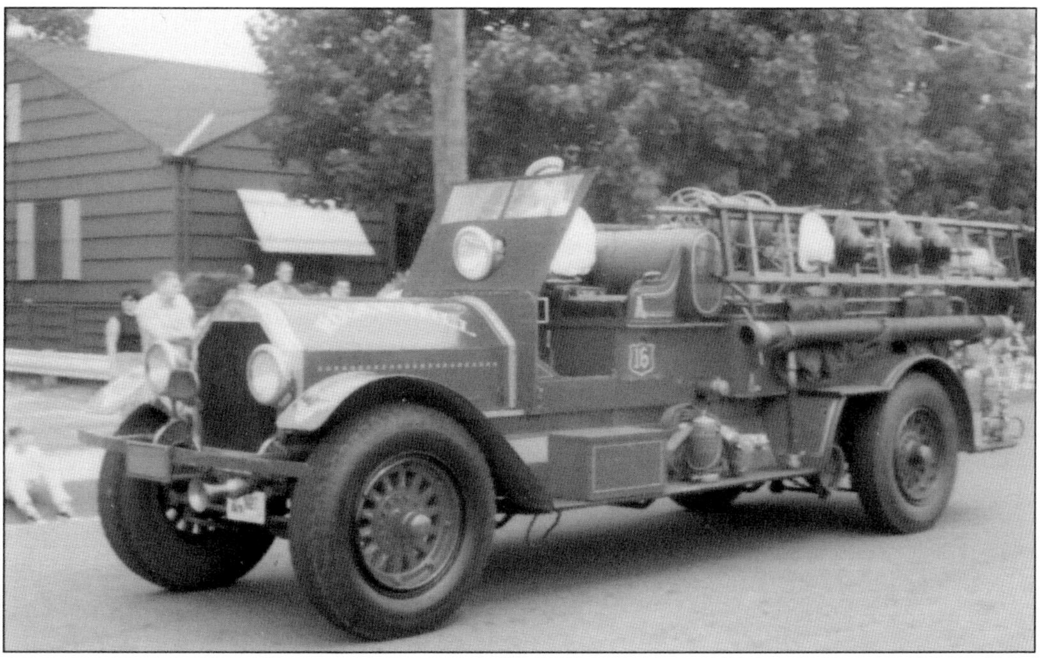

In the 1970s, Rocky Hill, Connecticut, used this former Springfield 1916 Seagrave pumper as a parade unit. A collector in North Carolina now owns it. (Courtesy Geoffery Neilsen.)

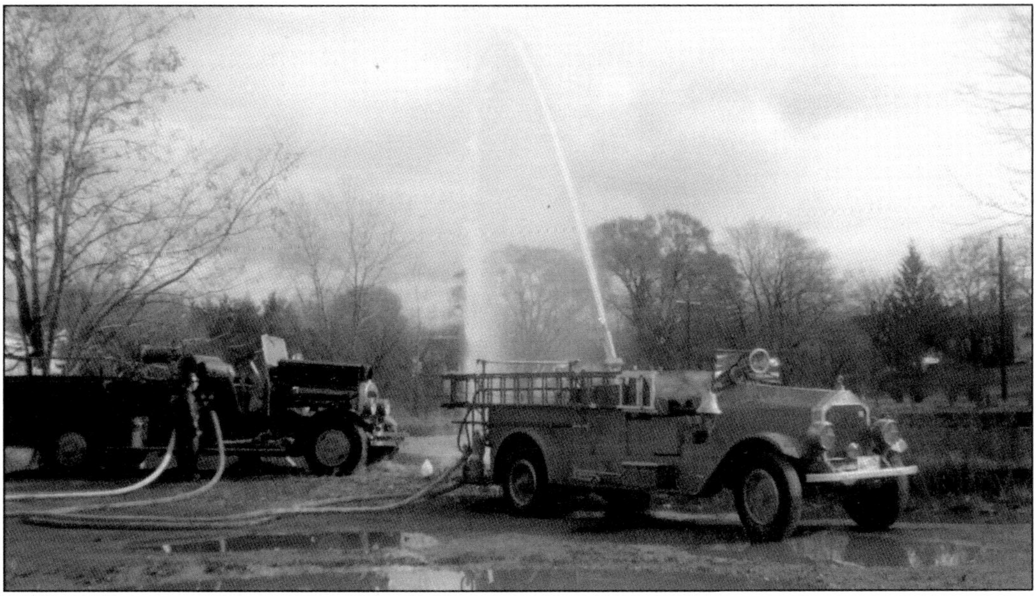

This 1975 view shows the former Springfield 1930 Seagrave pumper from the Connecticut Fire Museum participating in a drill with the Broad Brook Volunteer Fire Department of East Windsor, Connecticut. The 1930 Seagrave is drafting and pumping to a Broad Brook Fire Department truck that, in turn, is pumping into the museum's 1930 Maxim (formerly of Torrington, Connecticut) hose wagon equipped with a Hart wagon pipe formerly used on a Springfield Fire Department Mack hose wagon. (Courtesy Connecticut Fire Museum archives.)

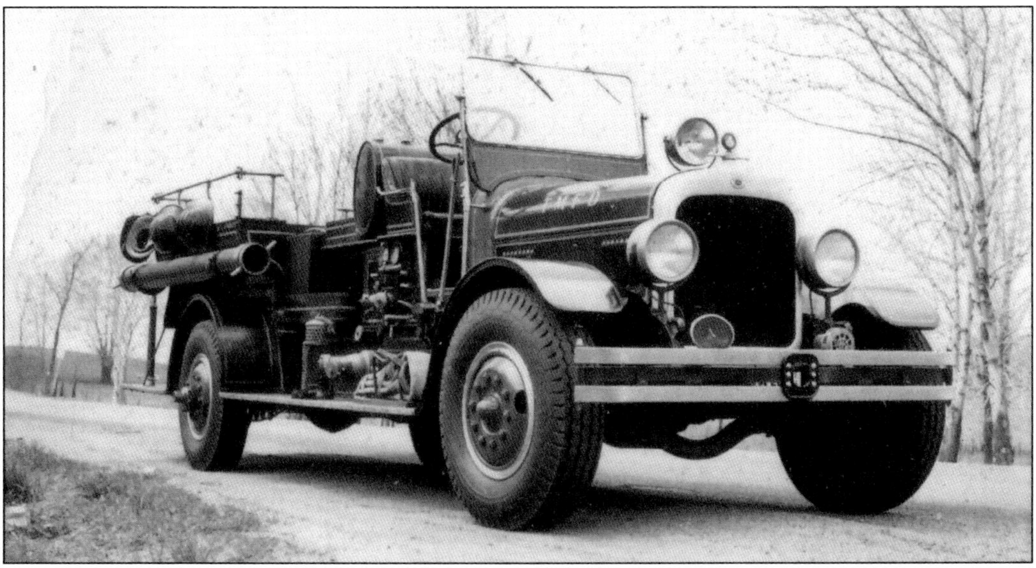

This is a former Springfield 1930 Seagrave 1,000-gallon-per-minute pumper purchased by the Connecticut Fire Museum. Because it was painted black, the firemen in Springfield referred to this truck as "Black Maria." It was the last black pumper in the department. It was placed in service on December 19, 1930, at Engine Company No. 8 in Winchester Square. At that time, it cost the city $13,300. The Seagrave also served at Engine Company No. 3 and Engine Company No. 16 until placed on standby service in June 1961. It was then used only as a spare to replace equipment out for repairs. (Courtesy Connecticut Fire Museum archives.)

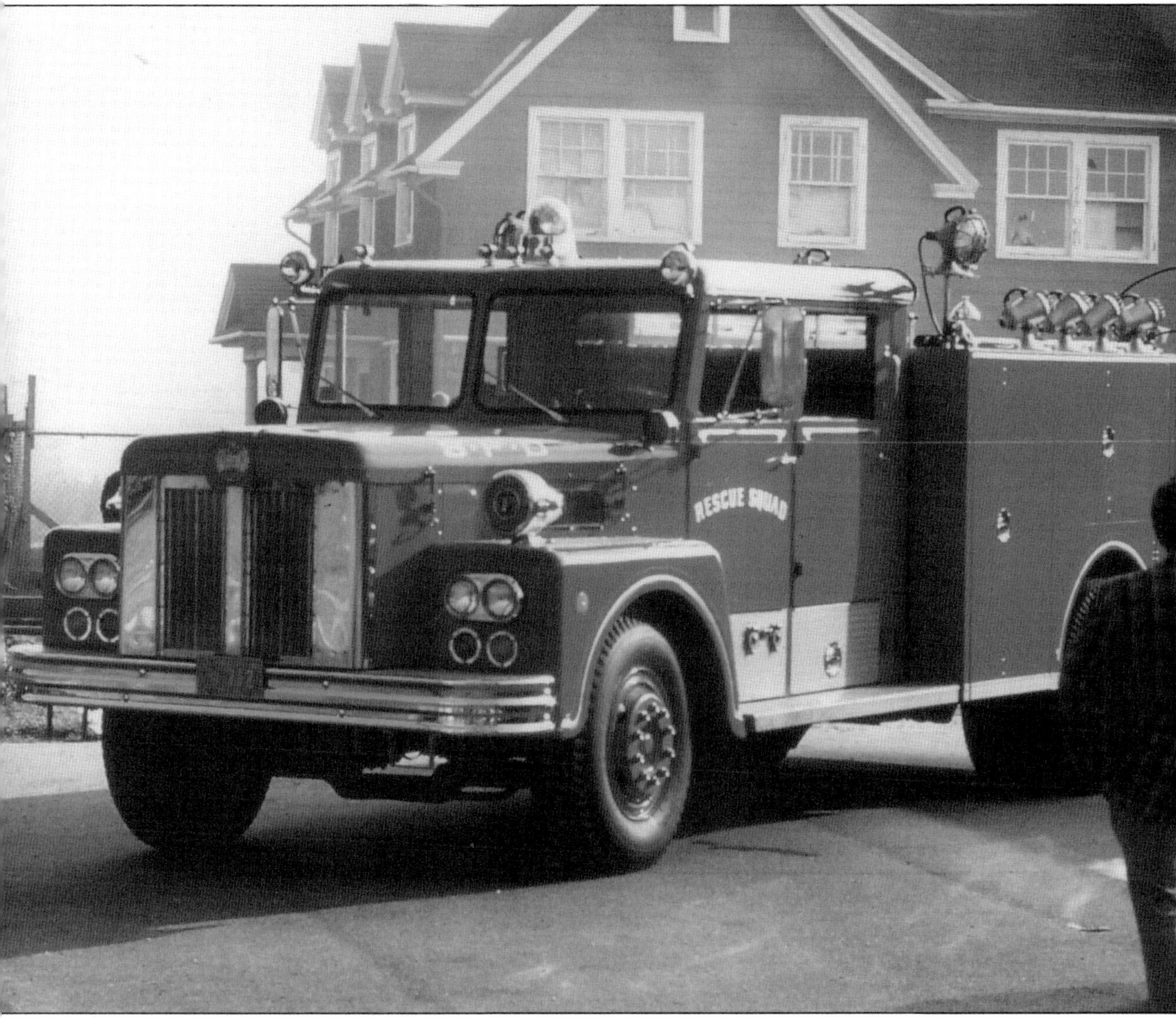

This 1973 Maxim rescue truck replaced the 1953 unit shown at the top of page 120. Springfield was always a loyal Maxim customer as long as it was in business. (Courtesy Geoffery Neilsen.)

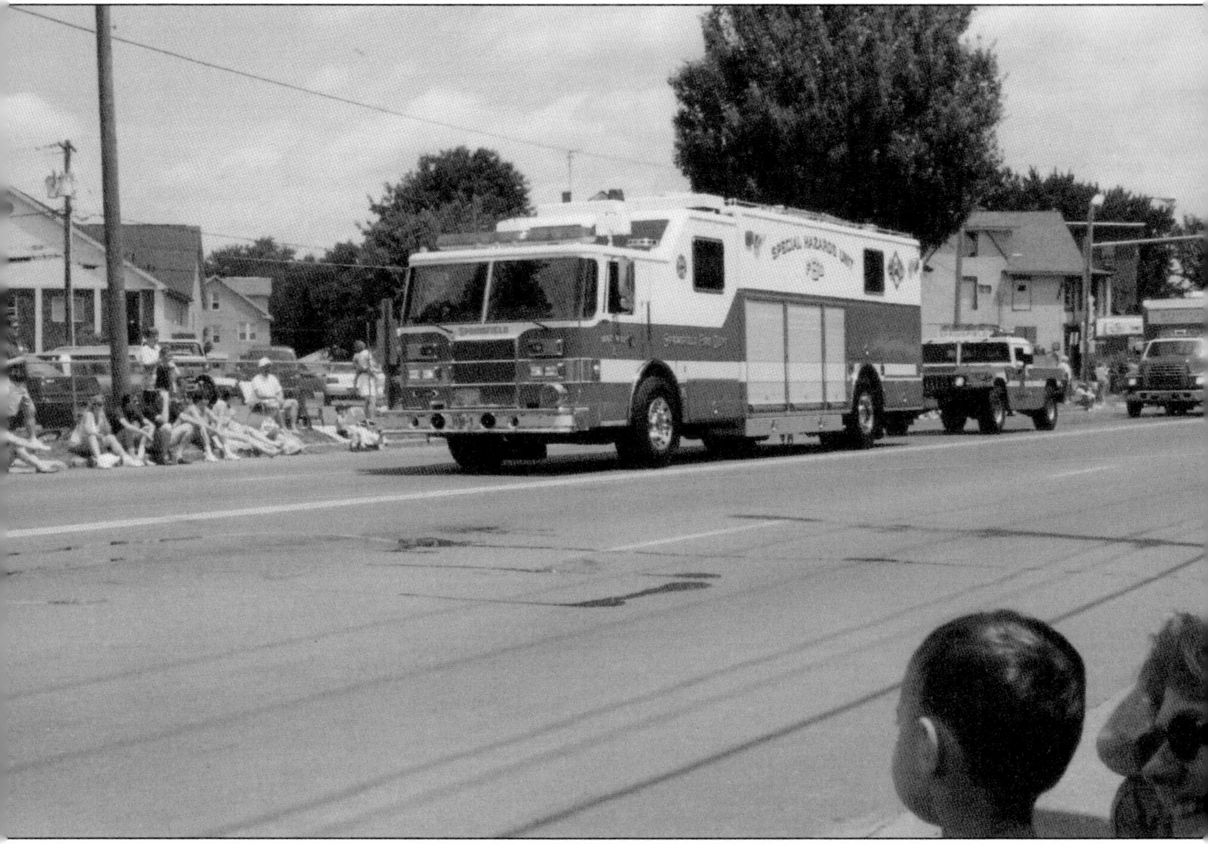

In 2006, Springfield ran this 1998 Pierce rescue truck, which replaced the 1973 unit shown on page 125. (Courtesy Roger Hildebrand.)

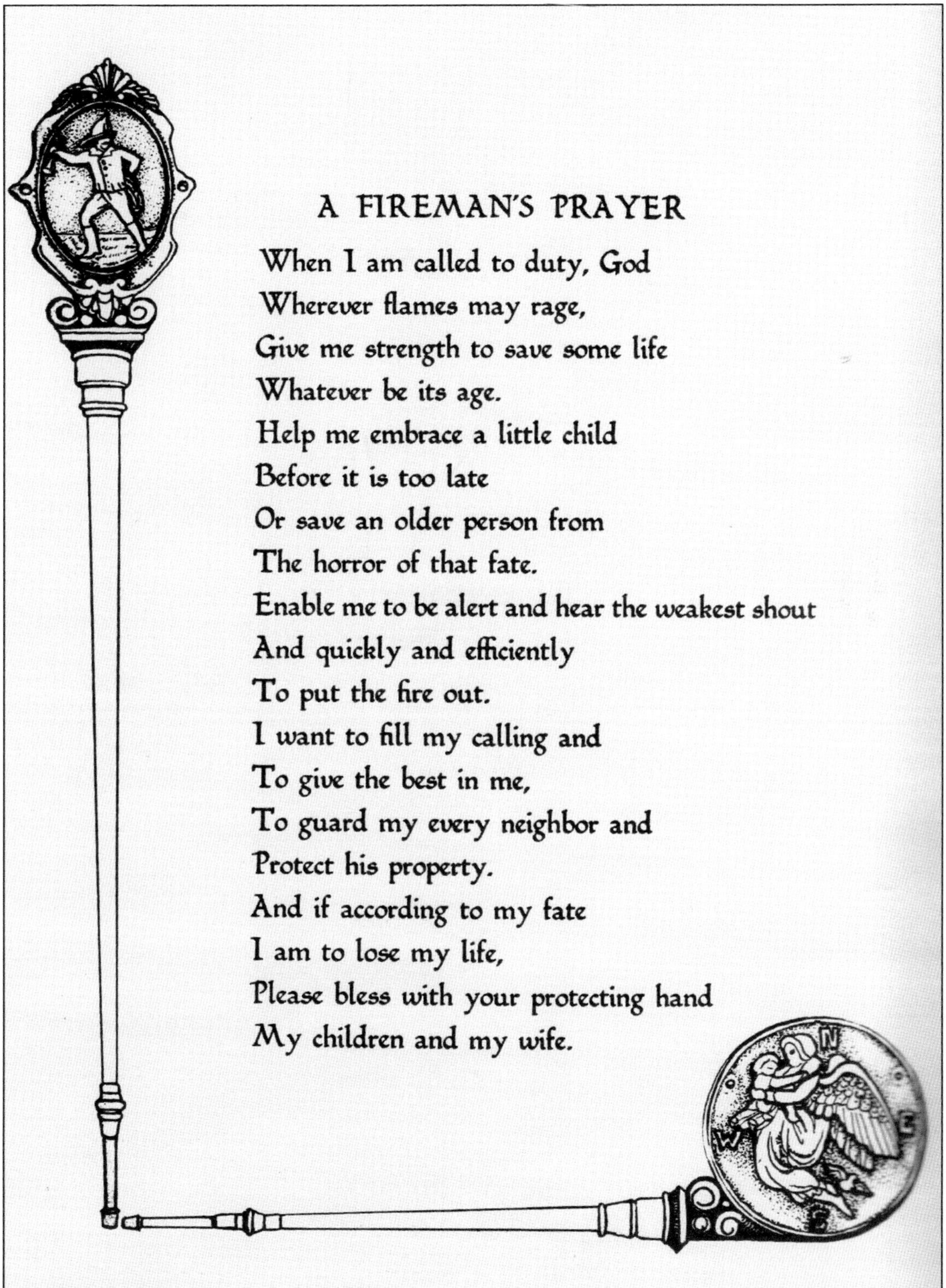

A FIREMAN'S PRAYER

When I am called to duty, God
Wherever flames may rage,
Give me strength to save some life
Whatever be its age.
Help me embrace a little child
Before it is too late
Or save an older person from
The horror of that fate.
Enable me to be alert and hear the weakest shout
And quickly and efficiently
To put the fire out.
I want to fill my calling and
To give the best in me,
To guard my every neighbor and
Protect his property.
And if according to my fate
I am to lose my life,
Please bless with your protecting hand
My children and my wife.

Pictured here is *A Fireman's Prayer.* (Courtesy Connecticut Fire Museum archives.)

ACROSS AMERICA, PEOPLE ARE DISCOVERING SOMETHING WONDERFUL. *THEIR HERITAGE.*

Arcadia Publishing is the leading local history publisher in the United States. With more than 3,000 titles in print and hundreds of new titles released every year, Arcadia has extensive specialized experience chronicling the history of communities and celebrating America's hidden stories, bringing to life the people, places, and events from the past. To discover the history of other communities across the nation, please visit:

www.arcadiapublishing.com

Customized search tools allow you to find regional history books about the town where you grew up, the cities where your friends and family live, the town where your parents met, or even that retirement spot you've been dreaming about.

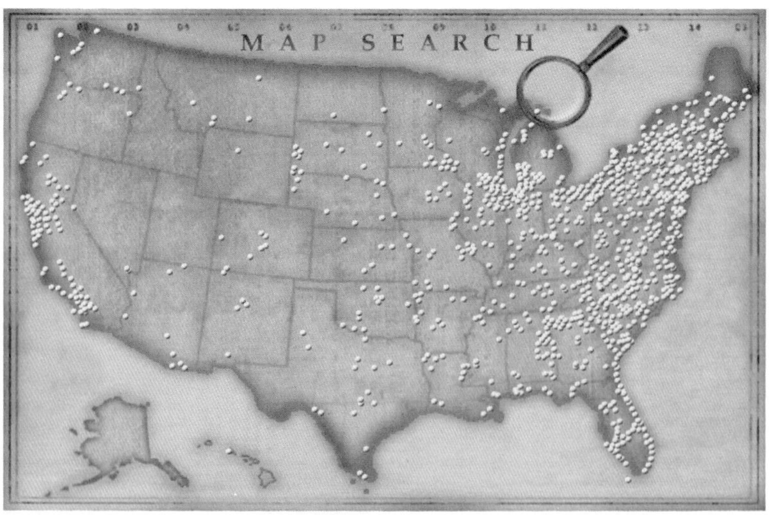